# What Every Engineer Should Know About Economic Decision Analysis

## WHAT EVERY ENGINEER SHOULD KNOW

A Series

*Editor*

**William H. Middendorf**

*Department of Electrical and Computer Engineering*
*University of Cincinnati*
*Cincinnati, Ohio*

**Vol. 1** What Every Engineer Should Know About Patents, *William G. Konold, Bruce Tittel, Donald F. Frei, and David S. Stallard*

**Vol. 2** What Every Engineer Should Know About Product Liability, *James F. Thorpe and William H. Middendorf*

**Vol. 3** What Every Engineer Should Know About Microcomputers: Hardware/Software Design: A Step-by-Step Example, *William S. Bennett and Carl F. Evert, Jr.*

**Vol. 4** What Every Engineer Should Know About Economic Decision Analysis, *Dean S. Shupe*

*Other volumes in preparation*

# What Every Engineer Should Know About Economic Decision Analysis

Dean S. Shupe
*Cincinnati, Ohio*

**MARCEL DEKKER, INC. New York and Basel**

**Library of Congress Cataloging in Publication Data**

Shupe, Dean, [Date]
What every engineer should know about economic decision analysis.

(What every engineer should know ; v.4)
Bibliography: p.
Includes index.
1. Engineering economy– Decision making. I. Title.
TA177.4.S5 658.1 80-22029
ISBN 0-8247-1019-3



Marcel Dekker, Inc.
270 Madison Avenue, New York, New York 10016

Current printing (last digit):
10 9 8 7 6 5 4 3 2 1

Printed in the United States of America

# Preface

Historians will record the decade of the 1980s as one of the most challenging and creative in humanity's short span, a period ushered in by our rude awakening to earth's resource limitations. Sudden and largely unexpected shifts in resource allocations and resource costs are stressing the fabric of our society, challenging our political and institutional structures, and threatening our quality of life.

Resource allocation and development decisions often hinge on technical and economic issues. These decisions bring crucial new responsibilities to the modern technologist—and a level of importance for the engineer that is unprecedented in peacetime.

This little book focuses on economic decision making. In the broadest sense, it addresses the issue of allocating scarce resources efficiently and more productively by examining the relative dollar values of alternative courses of action. Its unique objective is to present the essence of economic decision making in easily assimilated form without compromising theoretical soundness. Its intent is to provide a clear, concise, and perhaps even *fun* learning primer as well as a working manual for engineers, scientists, corporate management, government leaders, and private individuals involved in long-term investment decisions.

The book concentrates on a few essential principles. It weaves these principles through carefully selected, relevant examples, which progress steadily in difficulty as the complexities of an economic analysis are added in easily assimilated steps. Emphasis is placed on the systems approach so familiar to the technically trained wherein a decision-making entity is clearly defined and one's viewpoint clearly established. Cash flows occurring across the system boundary must be carefully identified for each alternative over the entire project life cycle. Differences in cash flow timings are dealt with through the use of interest. Cash flow diagrams are utilized for expediting visualization and analysis. While the cash flows are deliberately simple, the principles are applicable to the most complicated, irregular patterns encountered in practice.

Symbolism and terminology are in accordance with latest ANSI (American National Standards Institute) standards developed through the American Institute of Industrial Engineers. Extensive interest tables and equations are provided in the appendixes for easy access, together with a list of selected references for further reading.

It is our hope that this book will help popularize the science of economic decision analysis by clarifying its methodology and sensitizing users to its strengths and weaknesses, thereby enhancing its proper use and diminishing its misuse in the allocation of scarce resources.

*Dean S. Shupe*

# Acknowledgments

Our appreciation extends

To hundreds of enthusiastic students and short-course participants for their encouragement in writing this book

To a host of engineering friends and clients for the sharing of their experiences and suggestions

To the work of Eugene Grant and Grant Ireson as taught so effectively by Gerald Fleisher during my sojourn in Palo Alto

To graphic artist Jay Yocis for his outstanding illustrations

To typists Rhonda Christman and Linda Wefer for their patience through numerous revisions

To Alma for her long suffering

# Contents

# About the Author

Dean Shupe is currently a professor with the department of Mechanical and Industrial Engineering at the University of Cincinnati. His effectiveness as a teacher over more than a decade of engineering education is evidenced by his recognition through college and department teaching awards—eight in ten years—all initiated by students. In addition, he has received the University's Presidential Award for Excellence, the Outstanding Engineering Educator of Ohio Award, and the Western Electric Fund Award. His professional society memberships include the American Institute of Industrial Engineers (AIIE), American Society of Mechanical Engineers (ASME), the National and Ohio Societies of Professional Engineers (NSPE/OSPE), and the American Society for Engineering Education (ASEE). He is an engineering graduate of Iowa State University, Stanford, and MIT, a licensed professional engineer in Kentucky and Ohio, and an active engineering consultant to municipalities and industry.

# What Every Engineer Should Know About Economic Decision Analysis

# 1 Economic Consequences of Engineering Decisions

*Executives do many things in addition to making decisions. But only executives make decisions. The first managerial skill is, therefore, the making of effective decisions.*

Peter F. Drucker, *Management: Tasks, Responsibilities, Practices,* Harper and Row, New York (1974)

## Tough Decisions with Limited Resources

The ability to act decisively, oftentimes under severe pressure and with insufficient information, is the salient characteristic of the successful executive. To the extent that you agree with Peter Drucker's statement on decision making, you identify yourself as an executive. Decisions are continually required of those whose lives are satisfying and contributive in today's complex society. Whether one's responsibility is corporate or governmental leadership, family financial planning, or simply managing one's personal affairs, our lives are brimming over with decisions—decisions that may be trivial or crucial, depressing or exhilarating, temporal or irreversible. Tough decisions shatter some of us while challenging and strengthening others, because tough decisions originate from the inherent conflicts of life.

A decision is a choice—a choice between alternatives. The necessity for choosing is a direct result of our limited resources. Decisions are easy if we have plenty of money, plenty of time, plenty of materials, and plenty of talent. But a common characteristic shared by all societal groups, whether familial, corporate, or governmental, is that they are all confronted with the problem of *limited resources.* And so we must choose—choose how best to allocate our limited resources in the most productive manner.

## Engineering Economic Analysis: What It Is and Is Not

Engineering economic analysis is a quantitative tool that assists the decision maker in allocating scarce resources, especially scarce capital. It is a technique that compares alternatives on the basis of their projected economic outcomes. The economic impact of each alternative is first expressed in terms of its projected cash flows over a desired time period. The alternatives are then compared quantitatively by taking into account the differences in timing of the cash flows.

Engineering economy is not cost estimating. While the ability to forecast costs and benefits is necessary to a successful analysis, it is not the essential feature. Nor is engineering economy the science of operations research. While the optimization of system effectiveness through mathematical modeling may be invaluable in dealing with economic risk, operations research is not the essential feature. Nor is engineering economy the classical social science of macro- or microeconomics as concerned with gross national product, supply/demand, elasticity, or product pricing. Rather, engineering economic analysis is a quantitative approach for comparing alternatives on the basis of their projected cash flows, properly accounting for the time value of money.

## The New Engineer: Increased Social and Economic Responsibility

Engineers occupy a unique role in the modern technological world. Engineering is the one profession that is trained in the application of physical sciences and mathematics for solving humanity's problems. Traditionally the engineering function has been to utilize efficiently our planet's fundamental resources—labor, materials, and energy—to fulfill the needs of humanity. But the objective function is no longer simply stated in terms of cost per unit output or productivity per work-hour. While the material needs of humanity remain important, the complex issues confronting the modern engineer deal with the impact of engineering decisions on the *quality* of life—a friendly environment, freedom of thought and action, relative security, and the pursuit of happiness.

Recent surveys of corporate management confirm the resurgence of engineers at the highest levels of corporate leadership. Increasingly, engineers find themselves in positions of corporate power and responsibility either as the key executives or as close advisers to those who must decide. And so engineers must develop the managerial skills that are necessary for effective decision making.

## Major Considerations in Decision Making

The foundation of the decision-making process is *judgment.* Human judgment is based ultimately on human opinions, not on facts. The determination of the essential facts is in itself an ongoing, opinionated process.

The atmosphere of the decision-making process is *conflict.* Conflicts are a natural outcome of limited resources and lead directly to the need for a decision. The skilled executive encourages and cultivates dissent. Dissension and conflict stimulate creative and imaginative alternatives, clarify the question, and help identify the important considerations influencing the issue.

A multitude of important considerations may influence the decision. An important factor frequently encountered in decision making and essential to management's budgeting of scarce capital is *investment economics.* The economics of an investment addresses the questions: Will this investment pay? Will the return on investment be adequate? Should the present consumption of capital be sacrificed for the prospect of future benefits? These are the questions on which engineering economy are focused.

In addition to these economic questions, other considerations may be important, perhaps crucial, to the issue at hand. They include

Technical: Are the alternatives technically feasible?

Legal: Does the law constrain the choice of certain alternatives?

Financial: Is there sufficient capital available, regardless of the attractiveness of the economics?

Political: Are any alternatives influenced, perhaps eliminated, by political realities?

Health and safety: Do hazards preclude certain choices?

Environmental: How do the environmental impacts differ for the alternatives?

Moral: Is the decision influenced by societal mores?

Ethical: Are there conflicts with one's conscience or one's God?

The typical decision involving a complex issue in today's technological society usually hinges on several of these major considerations.

## Engineering Economy: Its Great Strength and Weakness

The common denominator in engineering economic analysis is *money.* Money provides a basis for comparing alternatives quantitatively, on an equal footing. By expressing the outcome of an alternative in terms of dollars, one is able to establish a level of objectivity that can be exceedingly valuable in decision making.

The principles of engineering economy are relatively simple and straightforward: defining one's viewpoint or system, identifying the cash flows to and from the system, focusing on differences in alternatives, and comparing the cash flows over equal time periods through interest manipulations. These principles are easily understood and committed to memory, if one has a good reference or tutor and a modest amount of practice. The challenge of engineering economy is in its application to conflicts. Formulation of the question to be decided and of the alternatives to be considered is critical to an effective decision. Too often the executive arrives at the right answer to the wrong question. Defining the appropriate alternatives and projecting their benefits and costs in terms of dollars require mature judgment in addition to a knowledge of engineering economic principles.

In evaluating air bags and other passive restraint devices for automobiles, what value should be placed on a human life? In establishing energy standards for office buildings, how does one assess the impact on work force productivity? In declaring war on cancer, what other human maladies must suffer from the resulting shift of medical resources? In mandating air pollution standards for automobiles and electric utilities, how much of the cost borne by society at large should be shifted to the consumer?

These are tough questions. They illustrate the essentially judgmental quality of decision making in an atmosphere of conflict. The principles of engineering economy may be very valuable in establishing the proper executive perspective necessary for an effective decision. In fact, a knowledge of engineering economy is absolutely essential if the issue is dominated by economic considerations. But if noneconomic factors are important, factors that are inexpressible in terms of dollars, then engineering economy alone is inadequate. For its great strength—the ability to compare alternatives quantitatively and objectively with money as the common denominator—is also its great weakness. When money fails as a quantifier for important considerations, then engineering economy fails as the primary decision tool and may lead to a false sense of objectivity.

The mature executive recognizes that judgment is the essence of decision making. The best decision is always an approximation. But in those instances where the economics dominate, where the crucial question is: Will this investment pay?—in those cases, engineering economy is the single most effective evaluation tool available to the decision maker today.

# 2

# Time Value of Money

## The Problem of Capital Budgeting

The productive allocation of capital is the pivotal function of top management in an enterprise. Through effective capital budgeting, the leadership of the enterprise prepares the foundation for its future profitability and vitality. The capital budgeting process provides the mechanism for rejuvenating or strengthening competitive abilities by shifting corporate resources from old, less productive areas of investment into new and more highly productive areas. It is in the arena of capital budgeting that engineering economy can perform such an effective and useful role.

Money has many uses. Two of its basic uses—to spend and to invest—exemplify the inherent tension that exists between present and future, a tension that one must deal with in the budgeting process. Invariably the alternatives differ in their time frames, just as present spending conflicts with future investing. If the alternatives are to be comparable, we must deal with this problem of timing differences.

## The Concept of Interest

The word *invest* embodies both present and future. To invest is to commit resources now in order to earn a return in the future. It is clearly implied that the future return will exceed the present investment. It is this anticipated growth in value that has enticed man to invest since the earliest moments of recorded history. If the resource is money, then this growth in value with time is called *interest.*

While the concept of interest is an old and familiar one, we nevertheless should proceed with some caution. The practical applications of the concept may be quite complex and will require a solid foundation that is clear and explicit. The concept of interest is the common thread running through the entire fabric of engineering economy. It provides the means for dealing with differences in timing.

If one invests \$100 now in exchange for the prospect of \$110 returned one year from now, then the \$10 increase in value is called the interest. Expressed in the customary percentage, the interest rate for the year is 10/100 or 10%.

The interest rate is the fulcrum that balances the present against the future. In the simple example above, \$100 now is balanced by \$110 at the end of 1 year, if the annual interest rate is 10%. Stated more concisely, \$100 now is equivalent to \$110 in 1 year at 10% interest. This statement can be expressed in equation form:

$$@\,10\%: \qquad \$100_{(\text{now})} = \$110_{(\text{in 1 year})}$$

The \$100 is sometimes referred to as the *present value* or *discount value* of the future \$110.

Increasing the interest rate shifts the balance toward the present. For example, if the interest rate were increased to 40%, the \$100 now would be equivalent to \$140 in one year. In the absence of the extra \$30, the balance would shift in favor of the present. This example illustrates a basic principle of economic analysis: *The higher the interest rate required for an investment, the more the balance favors the present, and the more difficult it becomes to justify an investment in the future.*

As a practical example of this important principle, consider an investment in energy-saving devices such as insulation for a home or manufacturing plant. The higher the time value of money (the interest rate), the more difficult it is to justify the investment of present dollars in exchange for future energy savings.

## Simple Versus Compound Interest

In the example above, a \$100 investment at 10% grows to \$110 in one year. If the investment were extended through a second year, the return at the end of the second year must obviously be greater than \$110. For *simple interest,* the return at the end of the second year is \$120, and is computed by multiplying the annual interest rate times 2 years. Simple interest is seldom used in the U.S. today.

For *compound interest,* the return at the end of the second year is \$121, and is computed by including the unpaid interest for the first year in the investment for the second year. Interest for the second year is \$11 or 10% of \$110, for a total return of \$121.

## Cash Flow/Time Diagrams

The ability to equate dollars at different moments in time via interest provides the essential mechanism for comparing economic alternatives on an equal footing. In contrast to the simple examples just examined, most practical alternatives involve the coming and going of multiple dollar streams at many different moments in time. To expedite the required interest calculations, a simple graphical technique is available for presenting the multiple dollar streams with their timing in a concise and explicit manner. It is the *cash flow diagram.*

The cash flow diagram is a graph of cash flow (in the vertical direction) as a function of time (in the horizontal direction), as shown in Figure 2.1. The time axis flows to the right and is labeled at the end of each time period, for example, 1 for the end of period 1. Zero (0) denotes end of period zero (or the beginning of period 1) and usually corresponds to the present time. The *end-of-year convention* is widely followed in engineering economy and will be used in the remainder of this text.

Cash flows are represented by vertical arrows located along the time axis at the time at which they occur. The magnitude of each cash flow is normally labeled near its arrow, rather than attempting to draw the diagram to scale. The direction of the cash flow (income or outgo) is indicated by the direction of the arrow.

Several conventions on direction are widely used. The choice is really a matter of personal preference. Our convention will be to represent *incoming* cash flows by arrows *to* the time line and *outgoing* cash flows by arrows *away* from the time line. This convention permits one to visualize oneself as located on the time line with arrows coming in and going out.

The previous examples used to introduce the concept of interest may be easily represented by cash flow diagrams. For the 1-year investment at 10%, the diagram is shown in Figure 2.2, while the 2-year investment diagram is shown in Figure 2.3.

Both diagrams are constructed from the viewpoint of the investor, that is, present dollars are expended for future dollars. From the viewpoint of the investor's environment, each diagram is inverted, as illustrated by Figure 2.4.

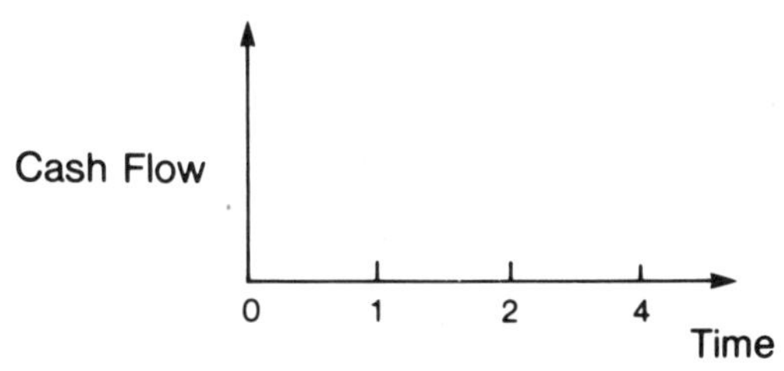


**Figure 2.1 The cash flow diagram**

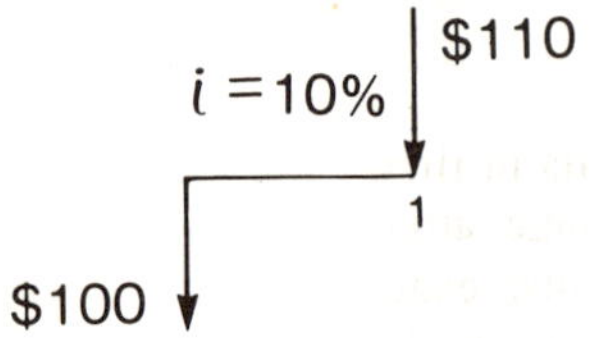


Figure 2.2 One-year loan earning 10% interest

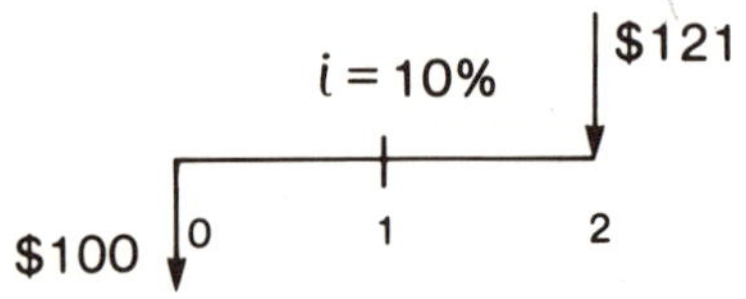


Figure 2.3 Two-year loan earning 10% interest (from lender's viewpoint)

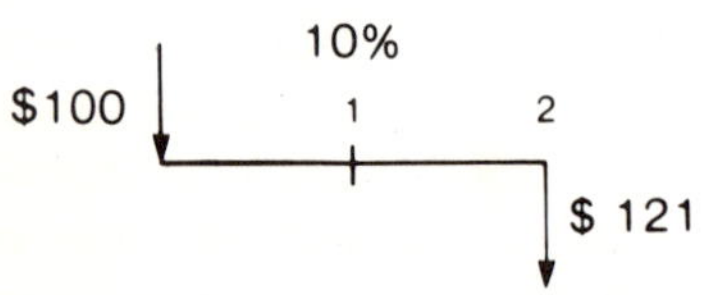


Figure 2.4 Two-year loan from borrower's viewpoint

The cash flow diagram is valuable in three important ways. First, it helps to clarify one's viewpoint for the economic analysis. In a sense, it requires a clear definition of the economic system. Second, it focuses attention on cash flow between the system and parties external to the system. Cash flows within subdivisions of the system are excluded. Likewise excluded are dollars that might not be actual cash flows, such as the accounting numbers for overhead and depreciation. Third, the cash flow diagram is an efficient and unequivocal method for communicating all cash flow information utilized in an economic analysis. As an information vehicle, it greatly reduces errors in interest computations during the analysis phase and facilitates the inspection and review of the data by all who must use the analysis results.

You are strongly urged to begin sketching and thinking in terms of cash flow diagrams. The cash flow diagram is a superb aid in visualizing and comparing alternatives. It expedites the necessary interest calculations, and it is simple.

### Whose Viewpoint?

Before moving on to the application of interest to several practical examples, we should comment briefly on the viewpoint for an analysis. The cash flow diagram shows cash transfers from a particular viewpoint, occurring to and from a particular economic system. Obviously, the choice of your system determines your resulting cash flows.

To illustrate how your viewpoint is important in defining your system, let us assume you are thinking about establishing a small business of your own. The business you are considering is the manufacture of cellulose insulation from recycled newspapers. Let us assume that the investment you need to start up the plant—purchase equipment, lease space, hire people, and market the product—is $200,000 including $40,000 of your own money and $160,000 as a bank loan. Of course you are keenly interested in whether the investment will pay. An economic analysis is essential. But whose viewpoint do you use? The bank's? Your own? Or both? If you choose yourself as the system, then you focus your attention on your own personal cash flows. Your $40,000 investment is obviously important. The payments on the loan are also important cash flows leaving "the system." In short, the appropriate cash flows consist of all dollars leaving or entering your own pockets.

You will probably also want to examine the profitability of the investment from the viewpoint of you and the bank combined. This combined viewpoint will be needed in convincing the bank to loan you the $160,000. Now the investment is the total of $200,000. Since the loan payments are between you and the bank, they are no longer important to the combined viewpoint and consequently do not show up on the cash flow diagram. The combined viewpoint is often referred to as that of 100% equity. When viewed through your own eyes, the profitability of the $40,000 investment is "levered" by the fixed loan obligation. Leverage can work for you or against you, depending on whether the interest rate actually realized by the manufacturing business exceeds or is less than the interest rate on the loan. The concept of leverage is a natural application of viewpoint to engineering economy and will be discussed in more detail in Chapter 7.

### Interest Factors and Equations

In order to express differing alternatives in commensurable economic terms, we will need a few basic interest equations together with a clear understanding of the conventions from which they are derived. Let us begin by recalling the $100 loan example used for introducing compound interest (Figure 2.3).

To find the $121, we computed the interest due at end of time period 1 ($10), added it to the initial $100, and then computed the interest due for period 2 based on the sum:

Year 1: $\$110 = \$100\,(1 + .10)$

Year 2: $\$121 = \$100\,(1 + .10)\,(1 + .10)$

If the present amount is called P and the future amount F, this period-to-period compounding may be represented by the general equation:

$$F = P\,(1 + i)^N \tag{1}$$

where i is the interest rate per compounding period and N is the number of compounding periods. *This is the underlying equation for all compound interest calculations.*

The equation can be rearranged into useful nondimensional ratios or *factors*:

$$\frac{F}{P} = (1 + i)^N \tag{2}$$

$$\frac{P}{F} = \frac{1}{(1 + i)^N} \tag{3}$$

P/F is often called the *present worth factor.* Other forms of the equations written explicitly for N and i are given in Appendix 1. Notice that F is defined as a single cash flow at the *end* of period N, while P is a single cash flow at the *beginning* of the first period. These definitions establish our conventions in agreement with ANSI standards,* and we must stick to them!

Now that we have the fundamental equation for compound interest under our belt, let us imagine that we are shopping for a 10-year, \$10,000 loan at 10%/year.

The skeleton cash flow diagram showing the loan without the payments is given in Figure 2.5.

Notice that the \$10,000 loan is shown as a cash flow into our pockets at time zero, while the repayment schedule is unspecified. As a matter of fact, several repayment schedules are widely used for loans of this type. Before we discuss the four most common schedules, you may want to think of them yourself and sketch their cash flow diagrams. The question is: How might a 10-year, \$10,000 loan be repaid?

One common repayment schedule is a single lump sum occurring at the end of the tenth year, as shown in Figure 2.6. The single repayment is typical of a

---

*ANSI Z94.5 *American National Standard Industrial Engineering Technology, Engineering Economy,* The American Society of Mechanical Engineers, New York (1972).

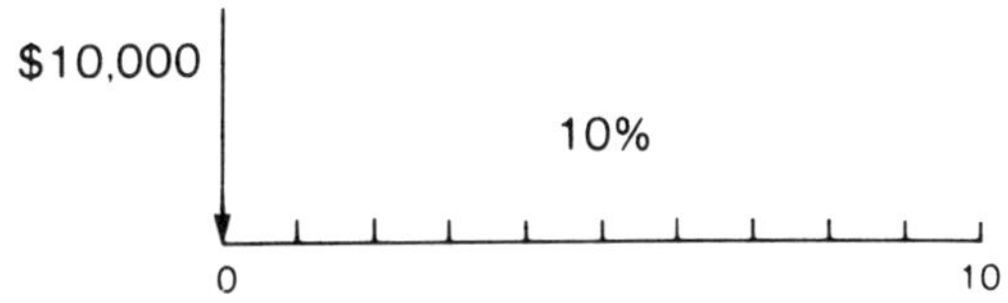


Figure 2.5 A $10,000 10-year loan with payment unspecified

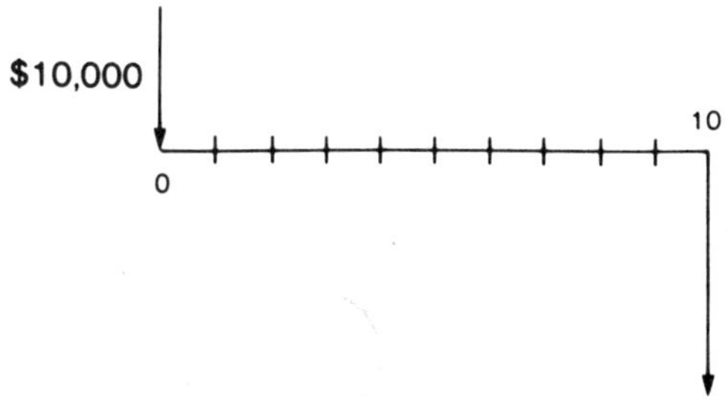


Figure 2.6 Single repayment of $10,000 loan

personal promissory note loan. But how large is the single payment? Perhaps you have already computed it by compounding the interest 10 times:

$$F = P(1 + i)^N = \$10{,}000\,(1.10)^{10}$$

$$= \$25{,}937$$

The resulting single payment required to pay off the loan at end of 10 years is $25,937. In other words, the future value of $10,000 at 10% interest is $25,937. Conversely, the present or discounted value of $25,937 at 10% interest is $10,000.

A second method of repaying the loan is by paying the interest at the end of each year, then paying the original loan amount or principal at the end of the tenth year. This repayment schedule is shown in Figure 2.7. Since the interest is paid as it accrues, the annual interest is a uniform $1000. Notice that in this case the sum of the repayments over the 10-year period is $20,000. This type of repayment schedule is characteristic of bonds, except that bonds normally pay interest semiannually.

A third method is to repay the principal in 10 equal, $1000-increments at the end of each year plus the interest due for that year. This repayment schedule

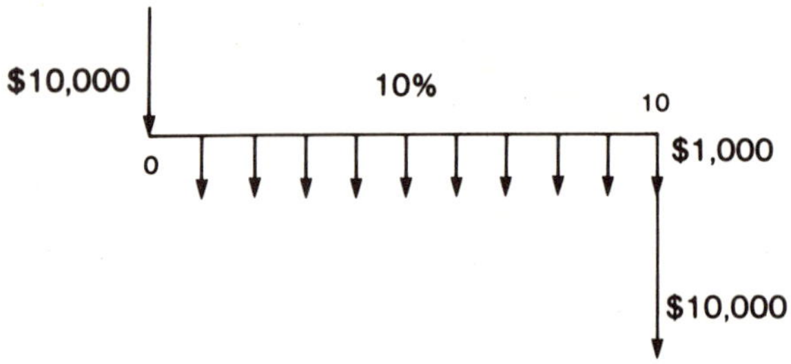


**Figure 2.7 Loan repaid by 10 equal annual interest payments and one principal payment**

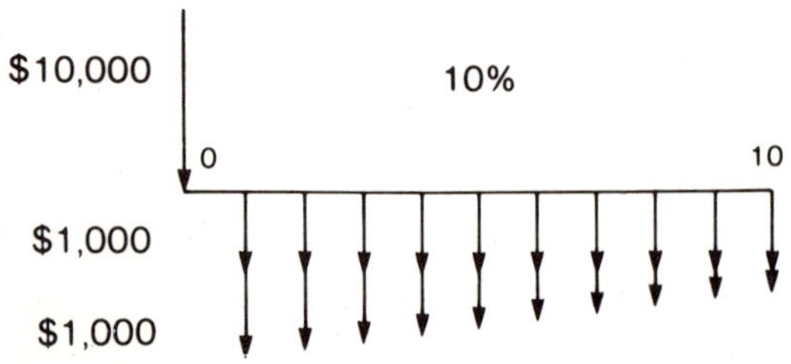


**Figure 2.8 Loan repaid by 10 equal annual principal payments plus interest on the unpaid balance**

is shown in Figure 2.8. The interest due at the end of the first year is 10% of the entire $10,000 or $1000. The second year's interest is 10% of $9000 or $900 (since $1000 of the principal has been repaid). Each year the payment decreases by $100 until the tenth year's repayment reaches $1100. What is the total repayment cash flow in this case? By summing, we can find it to be $15,500.

The fourth method of repayment commonly encountered is that of 10 uniform payments, as shown in Figure 2.9. This schedule is characteristic of home and auto mortgages except mortgage payments usually occur at the end of each month. In this case, it is not immediately obvious what the annual payment should be. As a matter of fact, it will require some work to determine the correct payment, but the results will be worth the effort.

Let us define the uniform end-of-period cash flow as A. The diagram may now be expressed in general terms of A (uniform payments), P (present amount), N (number of compounding periods), and i (interest per period), as shown in Figure 2.10. Again, we should clearly note that the convention for A is uniform payments occurring at the *end* of each period. We must adhere to this convention.

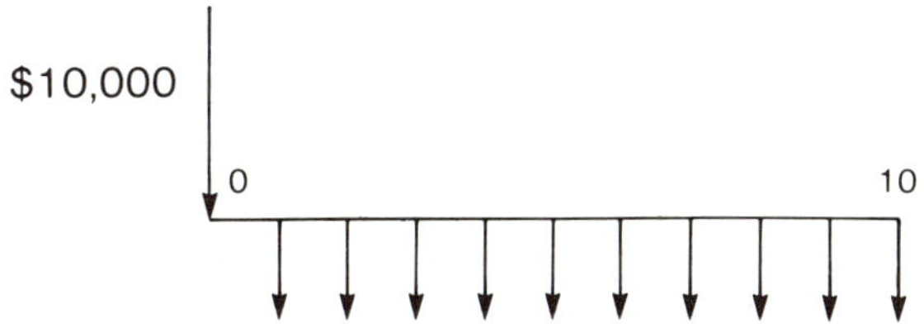


Figure 2.9 Loan repaid by 10 uniform annual payments

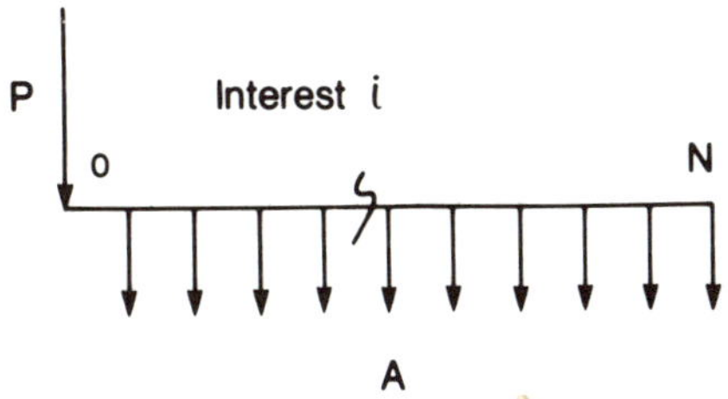


Figure 2.10 Standard conventions for P (present amount) and A (uniform payments)

Given P, N, and i, how do we find A? We will proceed by considering each A one at a time, beginning with the first year. Question: What is the present value of the first year's A? Using our P/F equation with N = 1, the answer is

$$P_1 = \frac{A}{1+i}$$

Now, what is the present value of the second A? This time we must move A back in time by 2 periods:

$$P_2 = \frac{A}{(1+i)^2}$$

If we continue this for all N years, and sum the results, then

$$\begin{aligned} P &= P_1 + P_2 + \cdots + P_N \\ &= \frac{A}{1+i} + \frac{A}{(1+i)^2} + \cdots + \frac{A}{(1+i)^N} \end{aligned}$$

This equation is cumbersome and unpleasant to use since it is a series of terms whose length varies with N. By manipulating the equation, we can convert it to a more useful format. First we divide by A and place the right side of the equation over a common denominator:

$$\frac{P}{A} = \frac{(1+i)^{N-1} + (1+i)^{N-2} + \cdots + 1}{(1+i)^N}$$

To shed the middle terms of the series, we multiply both sides by (1 + i):

$$\frac{P}{A}(1+i) = \frac{(1+i)^N + (1+i)^{N-1} + \cdots + (1+i)}{(1+i)^N}$$

and subtract the first equation from the second:

$$\frac{P}{A}i = \frac{(1+i)^N - 1}{(1+i)^N}$$

Dividing through by i, we obtain an explicit equation relating the present value P and the uniform payments A:

$$\frac{P}{A} = \frac{(1+i)^N - 1}{i(1+i)^N} \tag{4}$$

or

$$\frac{A}{P} = \frac{i(1+i)^N}{(1+i)^N - 1} \tag{5}$$

The ratio A/P is commonly called the *capital recovery factor.* An explicit equation for N is given in Appendix 1. The latter form of the equation may now be used to find the annual payment required for the $10,000 loan:

$$A = \frac{Pi(1+i)^N}{(1+i)^N - 1}$$

$$= \frac{(\$10{,}000)(.10)(1.10)^{10}}{(1.10)^{10} - 1} = \$1627$$

We find that an annual end-of-year payment of $1627 will pay off the $10,000 loan.

Before we leave the $10,000 loan example, we should notice the wide differences in the total cash needed to repay each of the four loans. And yet each repayment schedule is equivalent in its own way to $10,000 in present dollars. How can $25,937 for the first loan be equivalent to $20,000 for the second, $15,500 for the third, and $16,270 for the fourth? The answer is that the four loans differ in their *timing*. At a 10% time value of money, they are equal.

To solve the \$10,000 loan example, we derived interest formulas for the factors P/F, F/P, P/A, and A/P. These nondimensional ratios can be manipulated algebraically just as any fraction can be. For example, to find the ratio A/F we can proceed with a step-by-step derivation as we did with A/P or we can simply multiply A/P times P/F:

$$\frac{A}{F} = \frac{A}{P} \times \frac{P}{F} = \left[\frac{i(1+i)^N}{(1+i)^N - 1}\right] \left[\frac{1}{(1+i)^N}\right]$$

The result is

$$\frac{A}{F} = \frac{i}{(1+i)^N - 1} \tag{6}$$

or

$$\frac{F}{A} = \frac{(1+i)^N - 1}{i} \tag{7}$$

The A/F ratio is commonly called the *sinking fund factor*. The equation is written explicitly for N in Appendix 1.

For your convenience later, we will summarize now the interest factor equations and their names.

| | | |
|---|---|---|
| Compound amount factor, single payment | $\frac{F}{P} = (1+i)^N$ | (2) |
| Present worth factor, single payment | $\frac{P}{F} = \frac{1}{(1+i)^N}$ | (3) |
| Present worth factor, uniform series | $\frac{P}{A} = \frac{(1+i)^N - 1}{i(1+i)^N}$ | (4) |
| Capital recovery factor | $\frac{A}{P} = \frac{i(1+i)^N}{(1+i)^N - 1}$ | (5) |
| Sinking fund factor | $\frac{A}{F} = \frac{i}{(1+i)^N - 1}$ | (6) |
| Compound amount factor, uniform series | $\frac{F}{A} = \frac{(1+i)^N - 1}{i}$ | (7) |

Solving these equations each time you need them is not a particularly pleasant task. Fortunately, there are several alternatives available to you, depending on your individual needs and circumstances. They include the following:

Solve the equations as the need arises.

Refer to solutions prepared by others and given in interest tables such as those in Appendix 2.

Use preprogrammed or programmable hand-held calculators. All manufacturers of programmable calculators provide interest programs in their software manuals. A number of preprogrammed calculators are also available on the market, with hard internal wiring, hard-wired insert packages, or magnetic card libraries. For convenience in your personal programming, additional interest equations are given in Appendix 1 that are expressed explicitly in terms of N and i.

Use mainframe time-sharing or batch computers, especially when dealing with complex cash flows that require iterative solutions for i or multiple solutions for sensitivity and analysis.

The following examples will be solved using the interest tables in Appendix 2.

*Example 1: Retirement Savings Program* Many people prepare for retirement through a monthly or annual savings program. If $1000 is put aside at the end of each year in an investment earning 8%, find the total savings accumulated after 40 years.

The cash flow diagram is shown in Figure 2.11. Since we have assumed that the savings occur uniformly at the end of each year, then the savings cash flow corresponds to the conventional pattern for A. To determine the equivalent

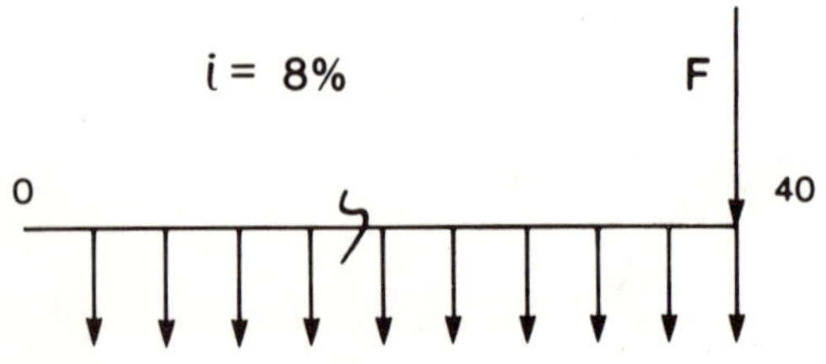


**Figure 2.11 Example 1: retirement savings program**

future amount, we use the F/A factor for i = 8% and N = 40 as given in Appendix 2:

$$F = \frac{F}{A}(\$1000) = (259.057)(\$1000) = \$259{,}057$$

The total savings of $40,000 grows to $259,057 through compounding over 40 years.

*Example 2: Inflation's Effect on Savings* While the savings in Example 1 appear impressive (over $1/4 million), the buying value of those dollars may shrink considerably due to inflation. If the average annual inflation rate during the 40 years of saving is 5%, find the value of the $259,057 in present dollars.

The cash flow diagrams for the $259,057 future amount and its equivalent present value P are shown in Figure 2.12. To find the equivalent present value P of the future amount F, we use the P/F factor:

$$P = \frac{P}{F}(259{,}057) = (0.1421)(\$259.057) = \$36{,}786$$

The P/F factor was obtained from the 5% table in Appendix 2. With 5% inflation, the value in present dollars of the 40 years of savings is much less impressive.

*Example 3: Retirement Annuity* The $259,057 savings accumulated after 40 years may be used to provide uniform annual payments as an *annuity*. For example, the savings may be withdrawn in uniform payments over the next 20 years (that is, years 41 through 60). Assuming that the savings are invested at 8% and that the annuity payments occur at the end of each year, find the annual payment.

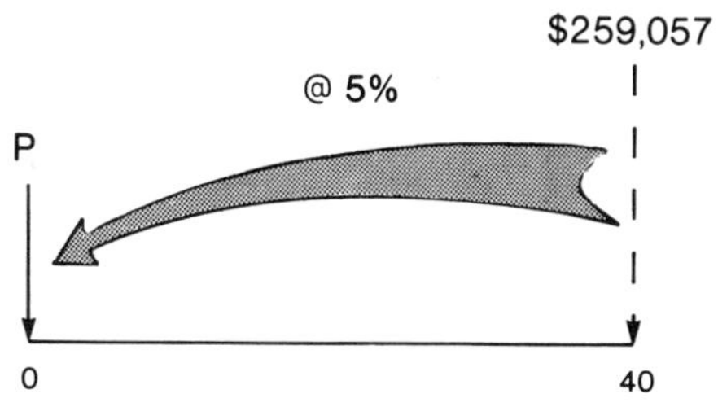


**Figure 2.12 Example 2: the equivalent value in present dollars of a future amount, using a 5% annual inflation rate.**

Now our time frame shifts to the fortieth year and the \$259,057 is now viewed as the present amount. The cash flow diagram for the \$259,057 and its equivalent uniform annual amount A are shown in Figure 2.13. Since the annuity payments occur uniformly at end-of-year, they correspond to the definition for A and we may therefore use the A/P factor for i = 8% and N = 20:

$$A = \frac{A}{P}(\$259{,}057) = (0.1019)(\$259{,}057) = \$26{,}398$$

This annuity continues for a fixed period of time, namely, 20 years. By sharing the risk of death with others through insurance, guaranteed annuities are also available for an uncertain time period such as the life of the insured or the insured and spouse.

*Example 4: Shrinkage of the Dollar* Historically, over the years the value of the dollar has shrunk due to inflation. How long does it take for the dollar's value to halve, if the annual inflation rate is 5%?

In this example, the number of periods N is unknown. Since the dollar's value is decreasing, \$2 in the future will have the buying power of \$1 now. The equivalent cash flow diagram is shown in Figure 2.14. We may use the F/P factor where:

$$\frac{F}{P} = 2$$

From the 5% table in Appendix 2, we find that N is between 14 and 15 years.

A useful rule of thumb for estimating the time required for an investment to double (or the dollar to halve) may be gotten by setting the F/P equation equal to 2 and solving for N over a range of interest rates:

$$\frac{F}{P} = (1 + i)^N = 2$$

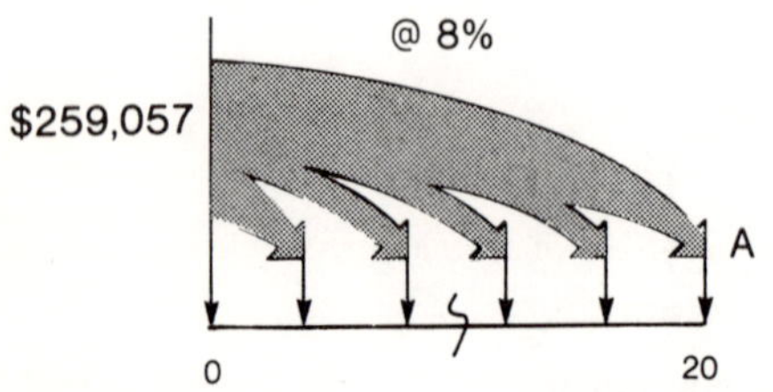


**Figure 2.13 Example 3: an equivalent retirement annuity from a present amount**

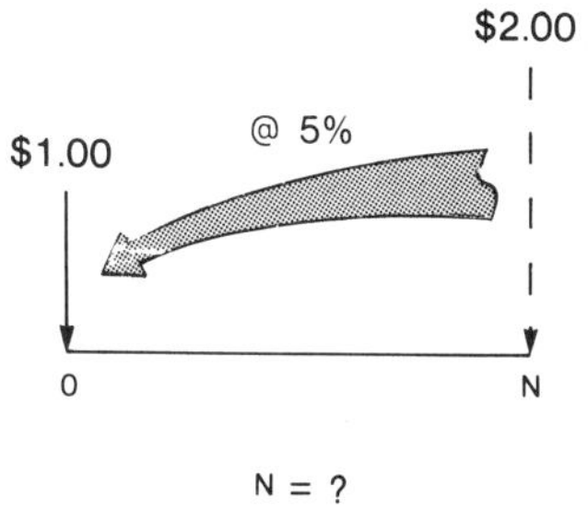


**Figure 2.14 Inflationary halving of a dollar's value**

Sample results are shown in Table 2.1. The right-hand column is the product of N times i where i is expressed in percent. The results indicate that the N × i product changes slowly over a relatively wide interest range. A resulting rule of thumb for the doubling of an investment is the *Rule of 74*, that is, N × i is approximately 74. For example, real estate that doubles in value in 7 years yields a return from market appreciation of approximately 74/7 or 10½%.

*Example 5: Ohio Sweepstakes Lottery* Eugene Hissner of Canton, Ohio recently won the $1 million sweepstakes in the Ohio lottery. Hissner, a former forklift driver in Canton, won $50,000 each year for 20 years. Find the equivalent present value of his winnings at an i of 10%, assuming the first payment occurs at the end of the first year.

The equivalent cash flow diagrams are shown in Figure 2.15. The end-of-year assumption regarding the annual payments corresponds to the convention for A. Therefore, we may use the P/A factor for 20 years at 10%:

$$P = \frac{P}{A}(\$50{,}000) = (8.5136)(\$50{,}000) = \$425{,}680$$

The present value at 10% is considerably less than $1 million.

**Table 2.1 Doubling periods N for an investment earning i% for period**

| i, % | N | N × i |
|---|---|---|
| 5 | 14.2 | 71.0 |
| 10 | 7.3 | 72.7 |
| 15 | 5.0 | 74.6 |
| 20 | 3.8 | 76.0 |
| 25 | 3.1 | 77.7 |

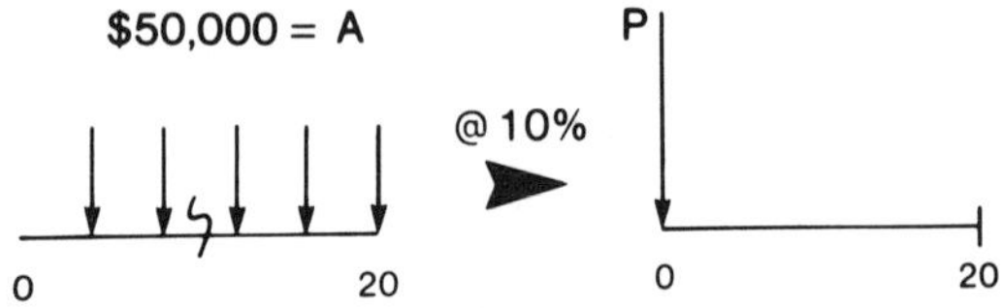


Figure 2.15 Example 5: equivalent present value of annual sweepstakes winnings (end of year)

The end-of-year assumption is unrealistic for sweepstake winnings, since the first payment is received soon after the drawing. Find the present value of his winnings if the first payment occurs at the *beginning* of the first year.

Figure 2.16 shows the cash flow diagram adjusted for the earlier payment. In this case, the cash flow diagram does not fit the A convention exactly so we cannot use the P/A factor for 20 years. However, by noticing that the first payment is already at the present time and the remaining 19 payments fit the A convention exactly, we may add payment 1 to the P value of the remaining 19 payments:

$$P = \$50{,}000 + \left(\frac{P}{A}\right)_{19} \$50{,}000 = \$50{,}000 + (8.3649)(\$50{,}000)$$

$$= \$468{,}250$$

The present value now is higher since Hissner receives each payment 1 year sooner. In fact, you might have noticed that the 1-year shift causes the value to increase by 10% (1.10 × \$425,680 = \$468,250). An important point of this example is that the conventions for A, P, and F must be followed *exactly*.

*Example 6: Life Insurance* The annual premium on a \$10,000 ordinary life insurance policy for a 25-year-old male is about \$150/year. An ordinary life policy guarantees a death benefit (\$10,000 in this case) in exchange for uniform annual premiums (\$150) paid during the lifetime of the insured. Using

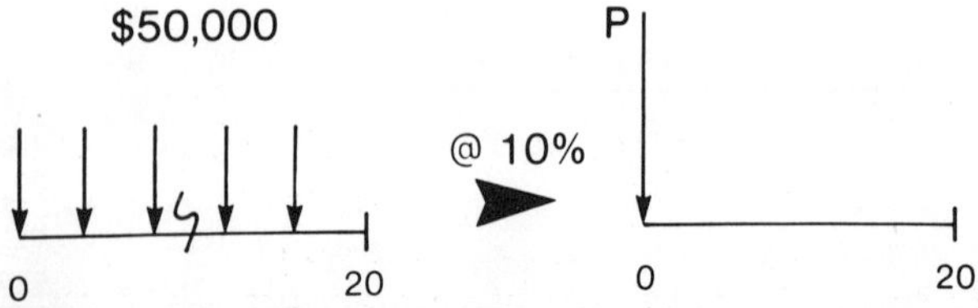


Figure 2.16 Equivalent present value of annual sweepstakes winnings received at beginning of each year

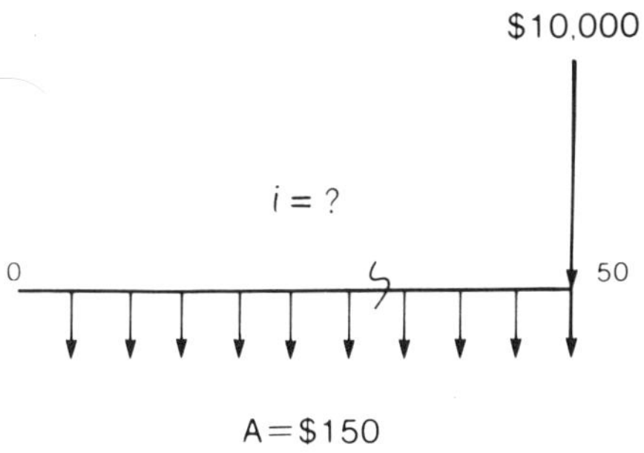


**Figure 2.17** Example 6: rate of return on a $10,000 whole life insurance policy

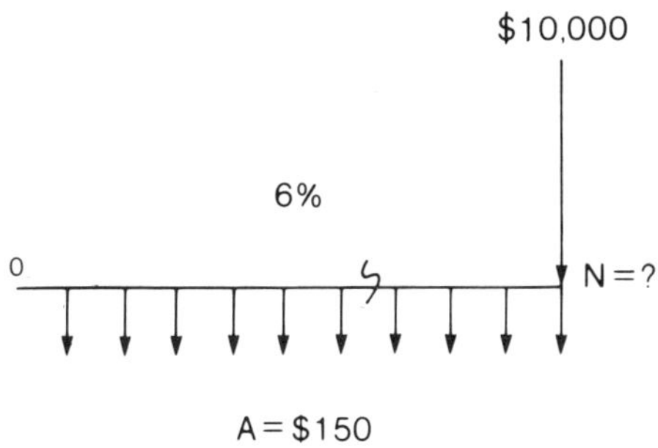


**Figure 2.18** Breakeven years N for a life insurance policy

the current life expectancy of 75 years for a male age 25, find the rate of return on the policy. Assume the payments are made at end-of-year.

Since end-of-year payments are assumed, the cash flow pattern fits the F/A pattern, as shown in Figure 2.17. Notice that we also assumed, unrealistically, that the fiftieth payment was made with receipt of the benefit. Since F and A are both known, we may solve for the F/A factor:

$$\frac{F}{A} = \frac{\$10,000}{\$150} = 66.7$$

Scanning the interest tables for N = 50, we find that F/A equals 66.7 for an interest rate between 1.0% and 1.25%. Consequently, the return on the life insurance policy for an individual living to age 75 is less than 1.25%.

As an alternative to life insurance, the 25-year-old man might invest the annual premium amount of $150 at a higher return rate and either gamble with an early death or purchase less expensive term insurance. If the annual $150 were invested at 6%, how long must the man live to break even with the $10,000 life insurance? Again assume end-of-year payments.

The cash flow diagram with i = 6% and unknown N is shown in Figure 2.18. The resulting F/A factor is 66.7 as before, but N is to be determined. From the 6% table in Appendix 2, we find that the breakeven N is between 27 and 28 years, or an age of 52 or 53.

*Example 7: Sinking Fund for Construction Bonds* In order to pay off its new construction bonds when they come due in 30 years, a local Massachusetts school district is required to deposit at the end of each year payments into a sinking fund account that earns 6% interest. Find the annual payment required to pay off $10 million in bonds in 30 years.

Since the cash flow diagram as shown in Figure 2.19 corresponds to our conventions for A and F, we will use the A/F or sinking fund factor:

$$A = \frac{A}{F}(\$10{,}000{,}000) = (.0127)(\$10{,}000{,}000) = \$127{,}000$$

*Example 8: Shenandoah Solar System* One of the largest U.S. structures to date to be both solar cooled and heated is the Shenandoah Community Center near Atlanta, Georgia. Total design and installation cost of the solar energy system was $726,000, with Energy Research and Development Administration (now the Department of Energy) providing funding. Annual savings were estimated as $77,000. For a system life of 20 years, find the expected return on investment.

This is the simplest classic investment example in which the present use of capital is foregone on behalf of anticipated savings in the future. If for simplicity the annual savings were assumed to occur at year's end (as shown in Figure 2.20), the pattern fits the P/A format:

$$\frac{P}{A} = \frac{\$726{,}000}{\$77{,}000} = 9.43$$

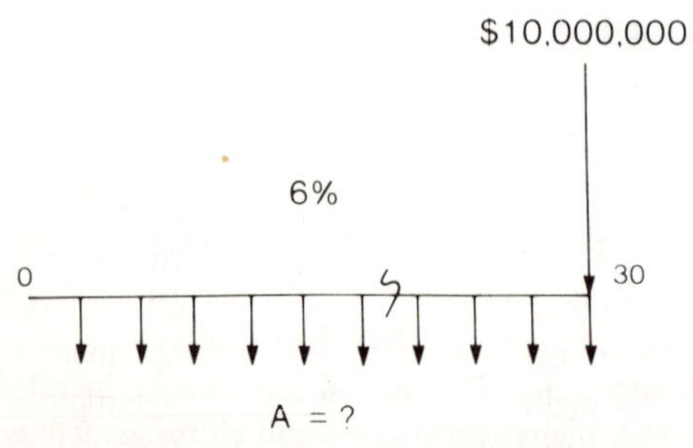


**Figure 2.19 Example 7: sinking fund**

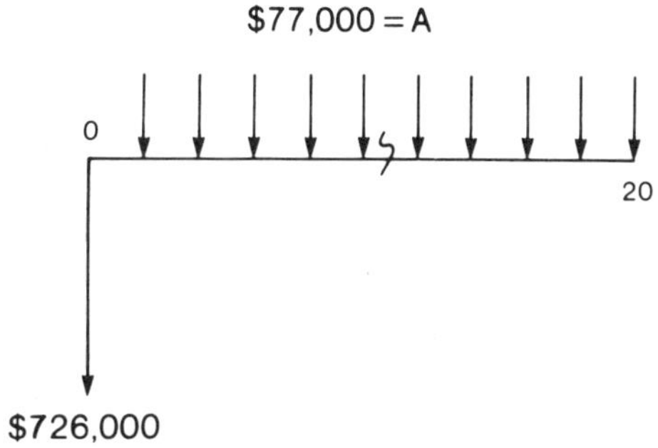


Figure 2.20 Example 8: Shenandoah solar system

Scanning the interest tables for a P/A value of 9.43 with N = 20, we find that the return i falls between 8% and 9%. If the fuel savings increase due to inflation in unit energy costs, the return would increase.

*Example 9: Rockwell Solid Waste Incinerator* Due to a recent winter natural gas shortage, Rockwell International's assembly plant in Marysville, Ohio nearly closed down. Following that crisis, the plant installed a $500,000 incinerator that burns the plant's combustible trash, thereby providing heating and cooling for the facility. Rockwell estimates the annual savings as $102,800 in electricity, $42,700 in natural gas, and $14,560 in trash disposal fees. Assuming an annual operating cost of $20,000 and a life of 10 years, find the incinerator's rate of return.

The multiple cash flows occurring each year, as shown in Figure 2.21, may be added algebraically to get a uniform net annual savings of $140,060. Assuming for simplicity that the savings occur at end-of-year, we may use either the A/P or P/A factor:

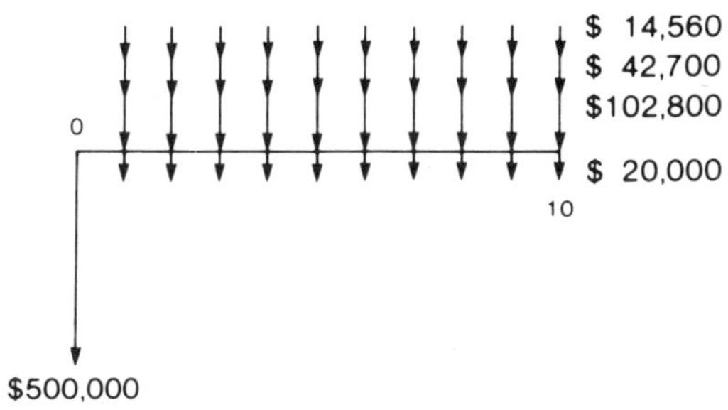


Figure 2.21 Example 9: Rockwell incinerator

$$\frac{A}{P} = \frac{\$140{,}060}{\$500{,}000} = .280$$

Scanning the tables for A/P (capital recovery factor) equal to .280 for N = 10, we find that the return on the investment is about 25%.

*Example 10: General Motors Citation* A fully equipped Chevrolet Citation may be purchased for about \$9000. If the anticipated salvage value at the end of 10 years is \$1000, find the equivalent annual cost of the invested capital, if the owner's interest rate is 12%.

This example is representative of a multitude of capital equipment investments involving a salvage value upon investment closeout, as well as first cost. Figure 2.22 illustrates the equivalent cash flow diagrams. The equivalent annual cost A is the difference between $A_1$ for the \$9000 and $A_2$ for the \$1000 salvage:

$$A = A_1 - A_2 = \frac{A}{P}(\$9000) - \frac{A}{F}(\$1000)$$

$$= (.1770)(\$9000) - (.0570)(\$1000)$$

$$= \$1593 - \$57 = \$1536$$

Notice that the salvage value is relatively unimportant because it occurs in the distant future. Also notice that if the auto is driven 10,000 miles each year, the equivalent annual cost of the capital apportioned on a per mile basis is \$1536/10,000 miles or \$.15/mile. Operating costs are additional.

## Effect of Compounding Frequency

Up to this point, we have assumed that interest is compounded annually. While it is customary to quote interest on an annual basis, compounding often occurs more frequently. Familiar examples include passbook savings accounts, credit cards, certificates of deposit, and home mortgages. We must recognize and deal with this minor complication before moving on.

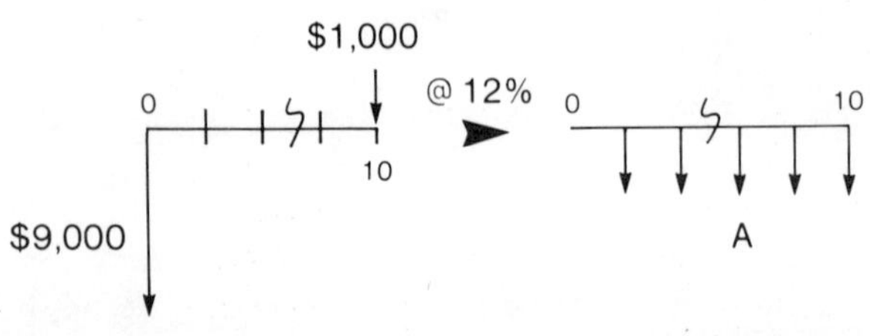


**Figure 2.22 Example 10: equivalent annual capital cost of automobile with anticipated salvage value**

Consider a gasoline credit card loan for which interest is charged each month on the unpaid balance. Since the interest is accumulated monthly, the compounding frequency is monthly or 12 times per year. Let us assume that we borrow \$100 at an interest of 1½%/month. Since it is customary to quote interest on an annual basis, an annual rate of 18% may be obtained by multiplying 1½ times 12 (the rate per compounding period times the number of periods per year). The resulting rate is called the *nominal annual rate r.* Actually the effective rate is somewhat higher, since interest is compounded during the year.

If the \$100 loan is permitted to ride an entire year without payment, the amount due after 12 months may be computed using our basic equation for compound interest by setting i equal to 1½%/month and compounding for 12 months:

$$F = \$100(1.015)^{12} = \$119.56 \tag{8}$$

At the end of one year, the total interest due is \$19.56 or 19.56% of \$100. Had the compounding frequency been only *once* each year, the annual rate would have had to be 19.56%. This can be formally shown by solving the basic equation for i with N = 1 (one compounding period):

$$F = \$100(1 + i) = \$119.56 \tag{9}$$

or

$$i = \frac{\$119.56}{\$100} - 1 = .1956$$

The rate based on annual compounding is called the *effective annual interest rate.* We have just shown that a *nominal* annual rate of 18% compounded 12 times per year is equivalent to an *effective* annual rate of 19.56%.

If the compounding fequency and the nominal rate are known, the effective annual rate i can be easily computed by generalizing and setting equal Eqs. (8) and (9):

$$\left(1 + \frac{r}{m}\right)^m = (1 + i) \tag{10}$$

where r is the nominal annual rate, m the compounding frequency, and i the effective annual rate.

Sample solutions for i are given in Appendix 3. Quarterly, monthly, weekly, daily, and continuous compounding are commonly encountered in the U.S. To compare alternatives having different compounding frequencies, we must first convert them to a common basis. The usual approach is to select annual compounding as the common basis and utilize effective annual rates for comparison.

Continuous compounding is the theoretical compounding limit. This case may be examined by taking the limit of Eq. (10):

$$\lim_{m \to \infty} \left(1 + \frac{r}{m}\right)^m = 1 + i \tag{11}$$

or

$$i = e^r - 1 \tag{12}$$

where the limit was recognized as a power of the natural log base e (2.7183).

Unfortunately, this theoretical expression for continuous compounding does not lend itself to practical situations where cash flows occur discretely throughout the year, for example, with savings accounts. Consequently, many financial institutions with continuous compounding have adopted a procedure that enables the computer to calculate interest 365 days of the year. Since federal law permits the use of 360 as the divisor in the compounding Eq. (10), the effective annual rate is actually somewhat greater than the theoretical limit:

$$i = \left(1 + \frac{r}{360}\right)^{365} - 1 \tag{13}$$

*Example 11: Eagle Savings* Eagle Savings Association pays \$6776 on a \$5000, 4-year certificate of deposit continuously compounded at a nominal rate of 7.50% (the legal limit for savings and loan institutions). Find the effective annual rate and compare to the theoretical limit.

Figure 2.23 shows the cash flow diagram for the viewpoint of the investor. To find the effective annual rate, we solve the basic interest equation, Eq. (1), for i:

$$\$6776 = \$5000\,(1 + i)^4$$

or

$$i = \left(\frac{\$6776}{\$5000}\right)^{1/4} - 1 = .0790$$

and

$$i = 7.90\%$$

The theoretical limit for continuous compounding may be found from Eq. (11) to be somewhat less:

$$i = e^r - 1 = e^{.075} - 1$$

$$= 7.79\%$$

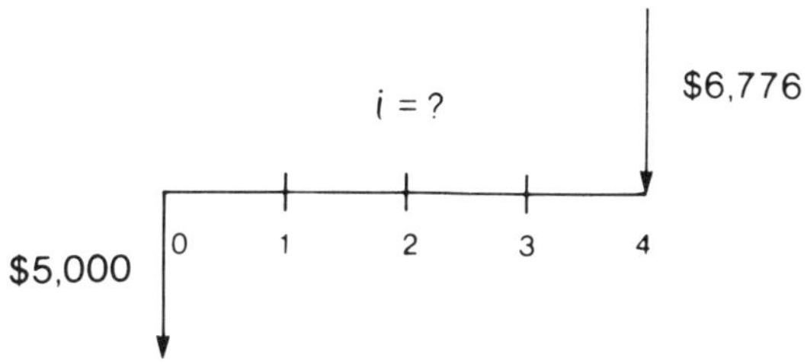


**Figure 2.23 Example 11: Eagle Savings 4-year certificate of deposit**

The effective rate is slightly greater than the theoretical limit for continuous compounding because Eagle's computer algorithm is based on the "365/360" equation:

$$i = \left(1 + \frac{.075}{360}\right)^{365} - 1$$

*Example 12: Cleveland Bank Loan* On the advice of his broker, a wealthy investor recently loaned \$100,000 at 10%/month to a Cleveland bank in financial trouble. Find the nominal and effective annual interest on the loan.

The nominal annual rate is simply the monthly rate times 12:

$$r = 12 \times 10\% = 120\%$$

The effective annual rate is found by compounding 12 times:

$$i = (1.10)^{12} - 1 = 214\%$$

*Example 13: Home Mortgage* On a 12% home mortgage for \$60,000 with monthly payments of \$700, find the amount of interest and principal in the first and second month's payments: Note: On home and auto mortgages, interest is usually compounded monthly, as the payments come due.

Consequently, the 12% is a nominal annual rate with interest compounded on the basis of 1%/month. The effective annual rate may be found from Eq. (10):

$$i = \left(1 + \frac{.10}{12}\right)^{12} - 1 = 12.68\%$$

The interest due at the end of the first month is 1% of the initial loan (\$60,000) or \$600 interest, with the remaining \$100 of the first payment applied to reduce the principal. The second month's interest is 1% of \$60,000 less \$100 (\$59,900) or \$599 interest, with \$101 applied to principal reduction.

The procedure continues through the life of the loan, with an increasing portion of the payment going toward the principal.

For a monthly payment of \$700, a 12% mortgage is paid off in about 16 years. If the payment is reduced to \$617, only \$17 of the first payment reduces the principal, and the life of the loan expands to about 30 years. A further reduction of only \$12 in the monthly payment (to \$605) extends the mortgage to 40 years!

***Example 14: Beneficial Finance Loan*** Recently, Beneficial Finance offered a 24-month, \$3000 personal loan for a monthly payment of \$158.61. Find the interest rate per month and the approximate nominal and effective annual rates.

The cash flow diagram from the viewpoint of the borrower is shown in Figure 2.24. The uniform end-of-period payments correspond to the A convention where the compounding period is a *month.* Therefore, we may use the P/A factor:

$$\frac{P}{A} = \frac{\$3000}{\$158.61} = 18.91$$

Scanning the tables for P/A = 18.91 and N = 24 compounding periods, we find that the rate per compounding period (month) is about 2%. At 2% per month, the nominal annual rate is 12 times 2% or 24%, while the corresponding effective annual rate is 26.8%:

$$i = (1.02)^{12} - 1 = 26.8\%$$

Notice that the interest tables are expressed in terms of N compounding periods with i the rate per compounding period. In this example, the compounding period is a month.

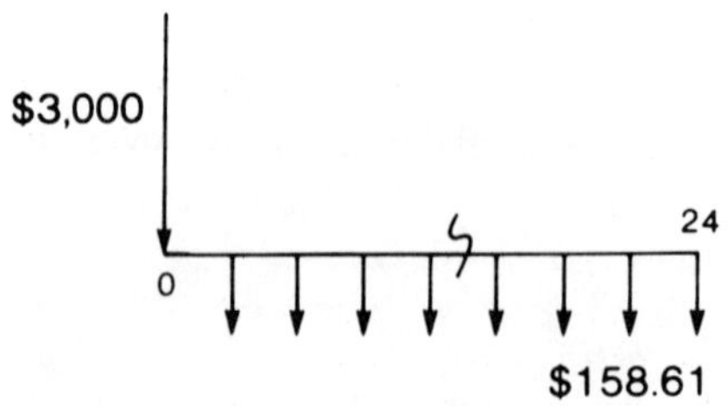


**Figure 2-24 Example 14: repayment of a 24-month loan**

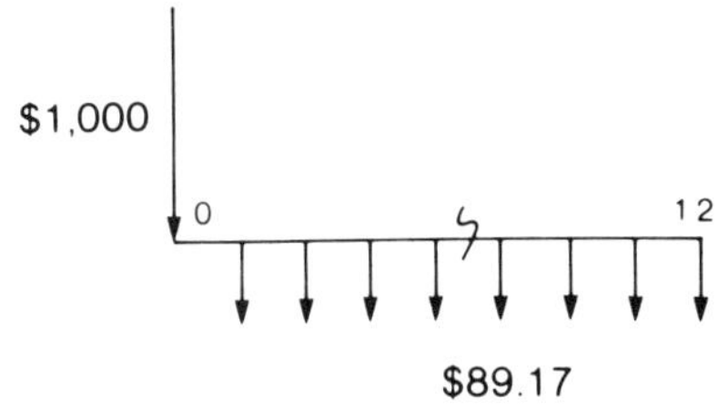


Figure 2.25 Example 15: a "7%" add-on auto loan

*Example 15: Add-on Auto Loan* A used car dealer offers you a "7% loan" for 12 months where the monthly payment on a $1000 loan is $89.17 ($1070/12). Find the nominal annual rate charged on the loan.

Figure 2.25 shows the cash flow diagram for the auto loan. It satisfies the P and A conventions:

$$\frac{P}{A} = \frac{\$1000}{\$89.17} = 11.2 \qquad \text{for } N = 12$$

From the interest tables, we find that the rate per month is somewhat greater than 1%, so the nominal rate exceeds 12%. Notice that the interest ($70) is computed on the initial loan of $1000. If you were allowed to pay off the loan in a single payment of $1070 at the end of 1 year, the effective rate would be 7%, but in this case you must begin payments after just 1 month. Consequently the effective rate is nearly double the quoted rate. The add-on loan is still encountered in the United States, although the federal truth-in-lending law requires written disclosure of the nominal and/or effective annual rates.

## Summary

Understanding the fundamental concepts introduced in this chapter will enable you to apply engineering economic analysis to a wide range of complex economic problems. Establishing one's viewpoint for the analysis, concentrating on cash flows over the life of each alternative as they occur for that particular viewpoint, developing the cash flow diagram as an efficient visualization and information tool, and using the concept of compound interest to equate cash flows at different moments in time—these are the four fundamental principles that undergird the successful and confident application of engineering economy.

# 3

# Differences Between Alternatives

Four approaches are commonly employed in economic decision analysis: (1) *present worth method*, (2) *annual cost method*, (3) *benefit/cost method*, and (4) *rate of return* or *return on investment method*. While sharing a common foundation of economic principles, these four methods differ in the ease with which they may be used to compare multiple interdependent alternatives. A discussion of project pairing and project independence is therefore a useful prelude to the detailed treatment of the four methods that follows in Chapter 4.

## Focusing on Differences

In all of the interest examples that we examined in Chapter 2, no mention was made of the alternatives. And yet the central act of decision making is the choosing among alternatives. Whether we are confronted with only 2 economic choices or 200, we must develop an approach for comparing rationally the alternatives on a commensurate basis.

In each of the preceding problems, only one alternative was stated explicitly. The second alternative was silently implied, namely, the *Do Nothing* alternative. To compare is to examine for similarities or differences. Similarities shared by all options are of little consequence to the one who must choose among them. The choice will hinge on *differences*. In the preceding problems, the cash flow diagrams were the cash flow differences between the explicitly stated alternative and Doing Nothing.

While a multitude of alternatives may be available, the actual comparison process is one of pairing the options and examining the differences by pairs. The words *challenger* and *defender* are often used to embolden this pairing process. By recognizing at the outset this inherent pairing characteristic, we will avoid the hazards that are associated with certain of the methods commonly employed for comparing multiple alternatives. In addition, by focusing on differences between alternatives we set the stage for assessing the degree of interdependence of alteratives.

*Example: Tippecanoe County Farmer* A retired farmer in Tippecanoe County, Indiana offers to sell you 10 acres of land on the outskirts of LaFayette for $60,000. As an added inducement, he offers to carry a loan at no interest for $10,000 cash down payment and $10,000 at the end of each year for 5 years. However, in the bargaining, he indicates that he would accept $50,000 cash in lieu of the loan. Find the effective annual interest that the farmer would be charging on the loan.

In this example, two options are stated explicitly: (I) to buy with a loan or (II) to buy with cash. The third alternative is implied, simply don't buy. The cash flow diagrams corresponding to the three alternatives are shown in Figure 3.1. To determine the interest rate charged on the loan, we compare the two purchase alternatives I and II.

What is the *difference* between this pair? In either case, the first year's cash flow will be at least $10,000. The question is: Should we pay out $40,000 more now or should we pay out $10,000 more a year for 5 years?

The cash flow difference is shown in Figure 3.2. This diagram may be recognized as simply the algebraic difference between alternatives I and II, more precisely, II minus I. Notice that in subtracting II minus I, the cash flow arrows

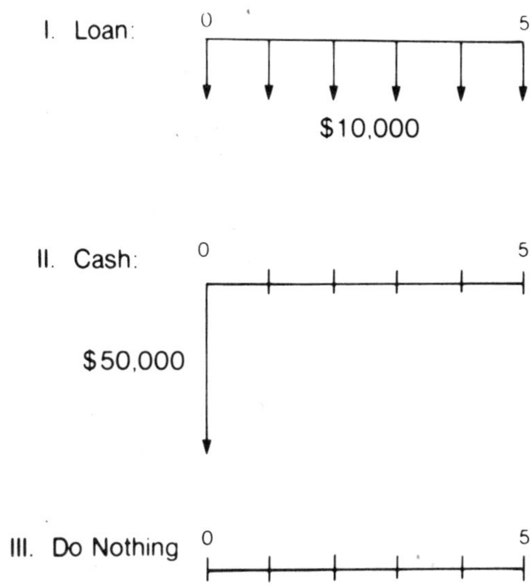


**Figure 3.1 Example: Tippecanoe farmer.**

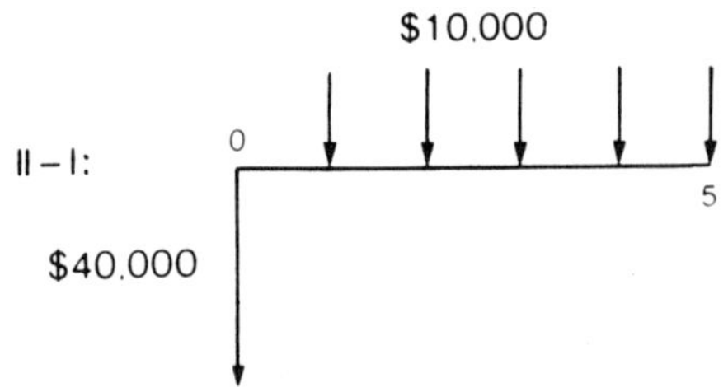


**Figure 3.2 Cash flow difference between cash purchase and 5-year loan**

for I are reversed from outgoing to incoming. The incoming arrows in Figure 3.2 represent the annual payments you avoid by paying cash. Cash payment results in $10,000 incoming cash flow difference a year, when compared to the loan.

To solve for i, we apply the P/A factor for 5 years:

$$\frac{P}{A} = \frac{\$40{,}000}{\$10{,}000} = 4$$

Scanning the tables for a P/A of 4 when N equals 5, we find that the farmer is charging about 8% interest on the loan.

This example illustrates the importance of cash flow differences in an economic analysis. The cash flow difference between a pair of alternatives may be determined by simply subtracting one cash flow diagram from the second. Regardless of the method selected for comparing alternatives, it is the cash flow difference that deserves our attention.

## Independent Alternatives

When top management allocates money for capital investments within a large corporation, the smaller projects under consideration are often *independent* of one another, that is to say, the selection of one project in no way influences the selection of any other.

As an example of project independence, let us consider a firm in which capital investment proposals of $50,000 or less are channeled up the chain of command for a decision at the vice-presidential level, for example, the vice-president of manufacturing. A list of projects in manufacturing might look like this:

Install process water recycling equipment.

Speed up packaging line.

Computerize the inventory system.

Enlarge warehouse facilities.

Install energy recovery devices on the boilers.

Insulate the roof of California plant.

Automate packing operation.

All seven of these projects are unrelated by application and therefore functionally independent. The \$50,000 ceiling identifies the projects as relatively "small" in the eyes of the firm, that is, projects for which sufficient funds are available for any or all that are economically acceptable. Since funds are not a constraint, the projects are financially independent.

As a result of their functional and financial independence, a comparison among these independent alternatives is inappropriate and incorrect. The appropriate comparison is between each proposed project and its challenger *do nothing.* Consequently, seven pairs of alternatives must be evaluated, with attention directed at the cash flow differences between each pair. The pairing of N independent alternatives is graphically depicted in Figure 3.3.

## Dependent Alternatives

Major investment projects, that is, projects that will require a substantial fraction of the total available capital, are invariably handled by top management on an individual basis. Because of the magnitude of the investment, the selection of one major project may very well preclude the selection of others. Due to sheer magnitude, the projects are financially dependent. For a small firm, an electronic typewriter or a new delivery van might be considered a major investment, while for a larger company, major projects might include a new manufacturing facility or development of a new product line.

In either case, projects that are financially dependent should be compared with one another in the decision process. If the selection of one project

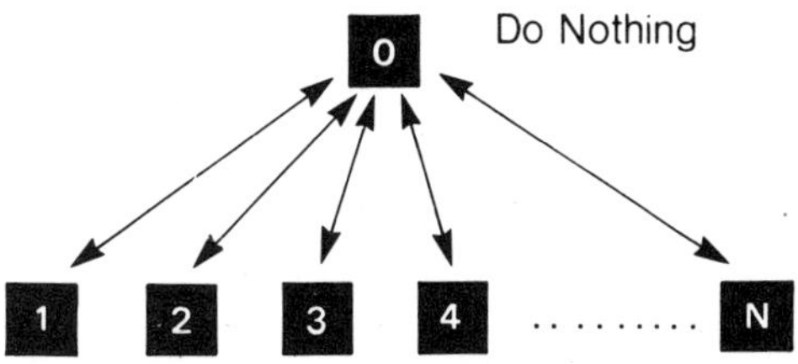


**Figure 3.3 Pairing of N independent alternatives**

precludes the selection of a second project, then the two projects are *mutually exclusive* and should be paired one on one. If there are, for example, four such mutually exclusive projects (that is, projects where choice of any one excludes all the others), then each project must be compared to each of the others. This pairing of four mutually exclusive alternatives is graphically illustrated in Figure 3.4.

Quite often, financially dependent alternatives are not mutually exclusive on a one-to-one basis. For example, let us assume that four projects are proposed for funding from a $100,000 pool of available capital:

| *Project* | *Cost* |
|---|---|
| A | $100,000 |
| B | $ 60,000 |
| C | $ 40,000 |
| D | $ 30,000 |
| Do Nothing | 0 |

Obviously the selection of project A excludes all other projects. But the other projects are smaller and may be accepted in different groupings that do not exceed the available funds. By arranging the projects into eight different groupings (A, B, C, D, B + C, B + D, C + D, and Do Nothing), the groupings may be treated as mutually exclusive alternatives.

In addition to financial dependency, many options are functionally dependent in that they fulfill the same need. Consider for example the municipality that has been directed by Environmental Protection Agency to either shut down its solid waste incinerator or bring it into compliance with federal air pollution

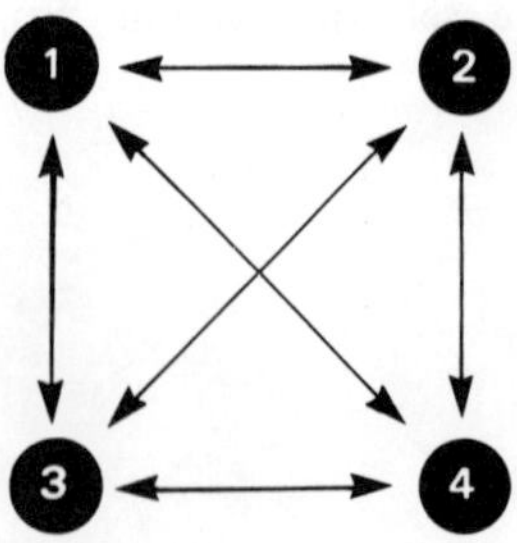


Figure 3.4 Pairing of four mutually exclusive alternatives

laws. The city is considering three primary options each with several suboptions:

1. Install electrostatic precipitators on existing incinerator:
    a. without energy recovery
    b. energy recovery on as-available basis
    c. energy recovery with backup oil boiler
2. Replace incinerator with resource recovery facility and recover:
    a. ferrous metals only
    b. all metals
    c. metals/glass/cellulose
    d. metals/glass/energy
3. Replace incinerator with a transfer station and landfill:
    a. landfill location A
    b. landfill location B

It is clear that the three primary options are mutually exclusive since any one will fulfill the existing need. The Do Nothing option is not available in this example because of legal constraints. Notice that the suboptions are contingent upon their respective primary options.

The alternatives may be depicted graphically as in Figure 3.5. Altogether we can identify nine alternatives that are mutually exclusive: 1A, 1B, 1C,

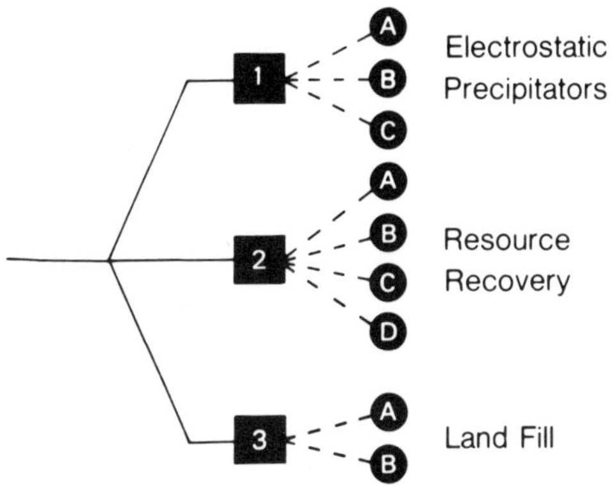


**Figure 3.5 Nine mutually exclusive alternatives for solid waste disposal**

2A, 2B, 2C, 2D, 3A, and 3B. In an economic analysis, each of these alternatives must be compared with each of the others.

Figure 3.4 illustrates the 6 pairings that occur for 4 mutually exclusive alternatives. Obviously the number of different combinations increases rapidly with the number of options available. It can be found from statistical mathematics that $N!/2(N - 2)!$ pair combinations exist for N items. For example, 36 pairs exist for 9 mutually exclusive alternatives—a result that you can confirm with patience by sketching a diagram in the manner of Figure 3.4.

The present worth and annual cost methods lend themselves readily to the economic comparison of mutually exclusive alternatives, whether there are only 2 or 200. The return on investment and benefit/cost methods are applied to multiple mutually exclusive alternatives only with some care and difficulty. Unfortunately the hazards accompanying these latter two methods are often unfamiliar to the novice.

# 4

# Life Cycle Comparisons

*The bitterness of costly operation lingers long after the sweetness of low first cost is forgotten.*

D. S. Shupe

## Life Cycle Cash Flows

Engineering economy is a common sense approach to capital budgeting. From its origin, it has recognized the need to compare cash flows over the entire project life rather than merely first cost. Its formal roots go back to 1887 when civilian engineer A. M. Wellington examined the trade-offs between first cost of railway location and follow-on operating costs.

Decision making in both the private and public sectors is improved when we look at the lifetime impact of all factors–downstream operating costs and savings as well as first costs. The "low bid" concept especially prevalent at one time in the public sector is being supplanted by "life cycle costing." Rapidly increasing energy and labor costs in times of continuing inflation make life cycle analysis absolutely essential in any significant economic decision.

In life cycle analysis, we examine the difference in cash flows that occur over the lives of the alternatives. But exactly what is the appropriate life to use? In some instances, the appropriate life is simply the physical life of an investment, that is, the time period during which the investment is physically capable of performing its intended function. As an example, a two-lane reinforced concrete bridge may survive for over 100 years.

Certainly the physical life represents an absolute upper limit. More often, the appropriate life for an economic analysis is the investment's useful life. The useful life, that is, the time period during which it performs its intended function efficiently and effectively, may be considerably shorter than its physical life, due to changing service requirements or technological obsolescence. An obsolete mainframe computer or a two-lane bridge carrying four-lane traffic are examples where the investment may have outlived its usefulness.

Occasionally neither the physical nor useful life is appropriate for the economic analysis but rather another period of concern to the particular investor. For example, if an engineering consultant is considering purchase of a professional office complex which he or she anticipates selling in 5 years, then the appropriate life cycle for examining the cash flow is simply 5 years.

## Comparing Rabbits and Rhinoceroses

In selecting the Grand Champion Angus bull at the Illinois State Fair each year, the breeding livestock judges compare the entrees on the basis of certain generally accepted standards. But if asked to select the Grand Champion bull from a mixed herd of Angus, Shorthorn, Hereford, and Charolais, the judges would throw up their hands and ask "On what basis?" Indeed, one cannot rank Angus and Shorthorn or rabbits and rhinoceroses without first establishing a common basis for comparison. Nor can one rank economic alternatives with differing life cycle cash flows without a common standard.

The strength of engineering economic analysis is that it provides a standard for the quantitative comparison of differing alternatives when the important differences are adequately represented by cash flows. As in the case of the \$10,000 loan where total repayment varied from \$25,937 for a single lump payment to \$15,500, we must deal with differences in *timing*. The concept of interest provides our common basis for comparison.

## Overview of Methods for Comparing Alternatives

Four formal methods involving the interest concept are widely used to compare multiple alternatives on a commensurate economic basis. Each of the methods requires selection of an interest rate that balances present and future cash flows.

The classic investment involves a trade-off of present expenditures in exchange for future benefits. Since the value of these future benefits when viewed from the present shrinks with increasing interest rate, selection of the proper interest rate is crucial to the outcome of the comparison. By increasing the interest rate, one decreases the importance of future dollars, thus eliminating all but those investments offering the highest return.

In an environment of limited capital, rationing may be accomplished by raising the interest rate to a suitable level. If an investment then fails to offer a return equal to or exceeding the established minimum, it is rejected. This minimum interest rate is often called the *minimum attractive return rate* and represented by the symbol $i^*$.

The four popular methods for comparing multiple alternatives are:

1. *Present worth method (PW)* Using the established value of i*, convert the cash flow for each alternative to its equivalent present value and compare these present values.
2. *Annual worth method (AW)* Using i*, convert the cash flow for each alternative to its equivalent uniform annual value and compare these annual values.
3. *Benefit/cost method (B/C)* Using i* and separating costs from the benefits, convert the cash flows to their equivalent annual (or present) values and compare the equivalent benefits to the equivalent costs for each alternative. Benefits must exceed costs.
4. *Return-on-investment method (ROI)* Find the interest rate i that balances present and future cash flows, and compare it to i*.

While all of these methods share a common foundation and yield comparable results, they differ in their suitability and ease of application to independent and dependent alternatives. We will point out those differences as we proceed. The fifth method—years-to-payback—is theoretically deficient but nonetheless very popular because of its directness and simplicity:

5. *Pseudomethod, years-to-payback* Find the number of years required for the future accumulated benefits to equal the initial cost.

## The Problem of Unequal Lives

Before we begin our discussion of the formal comparison methods, we must deal with the problem of unequal lives. We have seen that engineering economy is a decision-making tool for comparing multiple alternatives. The common denominator is money. The act of comparison is fundamentally a pairing process wherein the alternatives are examined two at a time. The focus is placed on differences in cash flow over the entire life cycle while handling timing differences via the interest concept. The appropriate life cycle or economic life is determined by the time frame of concern to us, as constrained by physical and useful lives.

As a general rule, *alternatives should always be compared over identical time periods.* It is patently unfair to compare project A with cheaper but shorter-lived project B without first asking what happens when B expires. If B must be replaced, then the replacement costs should be included in the analysis. If replacement is unnecessary due to a loss of the original need, then project A should be evaluated on the basis of the shorter life. In other words, the two projects should be compared on the basis of equal time frames, whether it is the shorter life of project B, the longer life of A, or some other appropriate time period.

In the case of N *independent* alternatives that have been channeled up the corporate structure for approval, each alternative is compared one-to-one against the option of Doing Nothing. It is a familiar fact that one can do nothing over any span of time—6 minutes, 6 months, or 60 years. Consequently the life of the Do Nothing alternative can always be adjusted to equal that of each of the other alternatives. Since the N alternatives are independent of each other, they need not and generally are not equal in life. For example, automating a packaging line may involve a life of 5 years while that of a warehouse enlargement may be 30 years. Since each is compared in turn against the option of Doing Nothing, no problem arises due to unequal lives.

It is in the case of *dependent* alternatives that we must consciously deal with the problem of unequal lives. For mutually exclusive alternatives, the choice is either/or, since selection of one alternative eliminates the others. For example, consider a city council about to authorize paving of a new subdivision. For the subdivision's traffic level, concrete is expected to serve about 30 years without major attention, while the physical life of bituminous paving is about 10 years. Although the concrete is initially more expensive, its maintenance costs will be lower. Since the two alternatives must be compared on the basis of equal time periods, a decision is needed regarding the appropriate period. In this case, council would probably select 30 years since the service requirement would extend with little change through the entire 30-year period. In other cases, technological changes or changing service needs could dictate use of a shorter time period.

If 30 years is selected as the economic life, then the cash flow diagram for the bituminous paving must be filled in for the entire 30 years. In other words, we must estimate the cost to repave the bituminous in 10 and 20 years.

For a more challenging example of unequal lives, consider a growing retail firm about to computerize its inventory, payroll, accounts payable/receivable, and management financial reports. One alternative is a minicomputer with vendor-supplied software that would be adequate for about 3 years. A larger, somewhat more flexible computer with in-house programming capability would probably serve for 5 years. What time period should be chosen for the economic analysis?

Some older textbooks recommend using the least common multiple, 15 years in this case, and assuming that each alternative repeats itself throughout the 15 years. This approach is inadequate during inflationary times, and particularly inappropriate in this case where the needs of the firm are growing rapidly and computer unit costs are decreasing due to technological advances.

A better approach is to predict the replacement for the smaller computer at the end of 3 years. Difficulty in making this prediction would suggest that the future uncertainties inherent in the decision favor using a 3-year economic life.

As you can see, the problem of unequal lives is not subject to simplistic solutions. A certain degree of experience and mature judgment is required to construct complete life cycle cash flow diagrams for the alternatives over identical time periods.

### Present Worth Method

In the PW method, the differing alternatives are made comparable by converting the cash flows for each into an equivalent present amount based on the minimum attractive return rate i*, and comparing these equivalent present worths. If the alternatives are mutually exclusive, they may be ranked according to their present worths. The best economic choice is the alternative with the highest present worth. If the alternatives are independent, then all alternatives with present worths greater than zero (corresponding ot the Do Nothing alternative) satisfy the i* criteria.

*Example 1: Dependent Alternatives (Solar System)* A solar system is under consideration for space and domestic hot water heating of a 900-square-foot, single-family residence in New York State. The annual heating cost for a conventional all-electric system for this modest residence is about \$1000. The cost estimates for three different sized solar system are shown in Table 4.1.

Our problem is to rank the four mutually exclusive alternatives (including alternative 0, no solar system) based on an owner's i*, for example, 10%. First we select a common economic life for all four alternatives. A life of 20 years appears appropriate, unless the owner plans to sell earlier. Next we sketch the cash flow diagram for each alternative and convert to its equivalent present worth using the appropriate interest factors or equations at 10%. The cash flow diagram for the smallest system A is shown in Figure 4.1.

System A's equivalent present worth is found to be \$554:

$$PW_A = -\$2000 + \$300\left(\frac{P}{A}\right) = -\$2000 + \$300\,(8.5136) = \$554$$

**Table 4.1 Installation data for three residential solar heating systems**

| System | A | B | C |
|---|---|---|---|
| Solar collector area, $ft^2$ | 100 | 200 | 500 |
| Storage tank, gal | 500 | 500 | 1000 |
| Extra cost of solar system | \$2000 | \$4000 | \$10,000 |
| Useful life, yr | 20 | 20 | 20 |
| Salvage value | 0 | 0 | \$1000 |
| Annual energy savings | 30% | 50% | 90% |

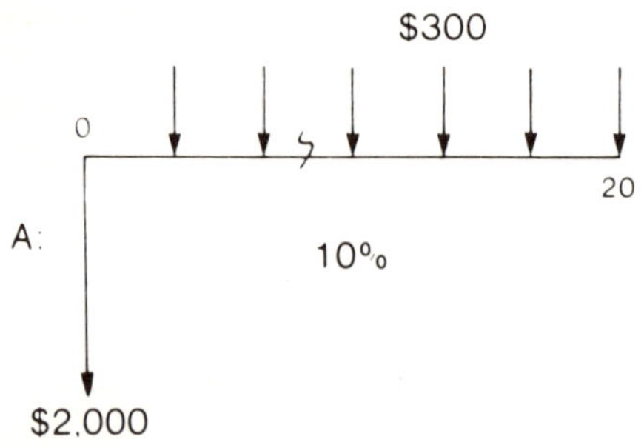


**Figure 4.1 Example 1: Solar system A (versus Do Nothing)**

For system B (Figure 4.2) the equivalent present worth is $257:

$$PW_B = -\$4000 + \$500\left(\frac{P}{A}\right) = -\$4000 + (\$500)(8.5136) = \$257$$

The present worth of system C (Figure 4.3) at an i* of 10% is a negative $2189:

$$PW_C = -\$10{,}000 + \$900\left(\frac{P}{A}\right) + \$1000\left(\frac{P}{F}\right)$$

$$= -\$10{,}000 + \$900\,(8.5136) + (\$100\,(.1487) = -\$2189$$

Finally we summarize the results by ranking the four options according to their equivalent present worths:

| *System* | *PW, $* |
|---|---|
| A | 554 |
| B | 257 |
| O | 0 |
| C | -2189 |

The best economic choice is system A while the worst is clearly C. Notice that the pairing process is accomplished automatically with the PW method. For example, the cash flow differences between options A and B are $2000 difference in first cost and $200 in annual savings. The present worth of the difference ($PW_B - PW_A$) is -$297 ($257 minus $554), in agreement with our ranking:

$$PW_B - PW_A = -\$2000 + \$200\,(8.5136) = \$297$$

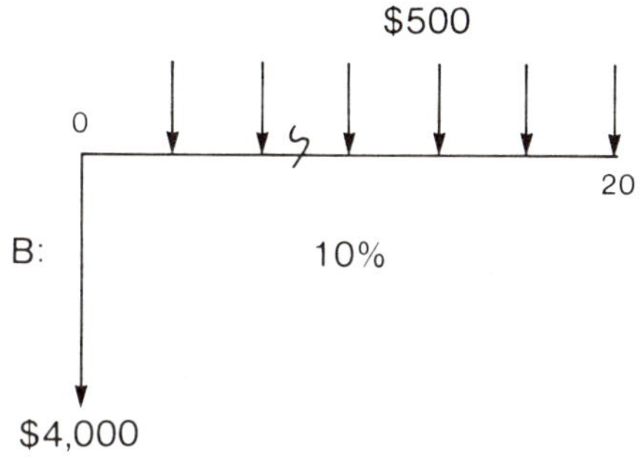


Figure 4.2 Example 1: Solar system B

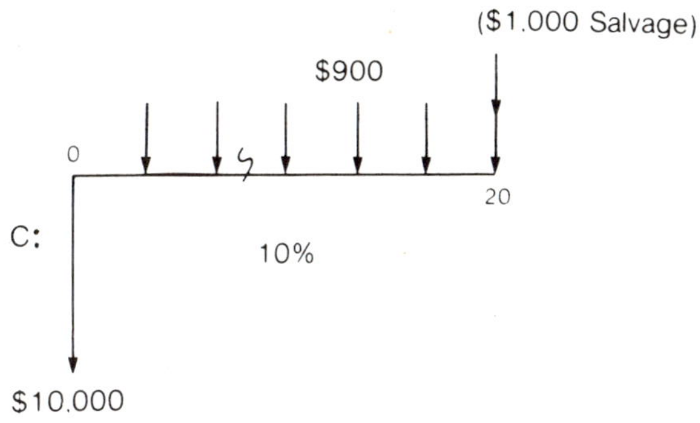


Figure 4.3 Example 1: Solar system C

This automatic pairing feature is a definite advantage of the PW method when dealing with multiple alternatives that are mutually exclusive.

In this solar example, fuel costs were assumed constant for simplicity. Escalating fuel costs (resulting in increasing energy savings) would tend to support investment in a larger solar system, while a higher i* value would support a smaller investment.

*Example 2: Independent Alternatives (TransAmerican Industries)* Four independent proposals have been submitted for consideration to the corporate budgeting committee of TransAmerican Industries, Inc. The four proposals are summarized in Table 4.2.

In this example, the proposals should not be compared to each other, since they are independent of one another. Any or all are acceptable. Unequal lives

**Table 4.2 TransAmerican Industries capital budgeting proposals**

| Project: | A | B | C | D |
|---|---|---|---|---|
| in \$1000s | Automate packaging line, \$ | Install heat recovery equipment, \$ | Computerize inventory system, \$ | Enlarge warehouse \$ |
| First cost | 200 | 1500 | 300 | 1600 |
| Service life, yr | 5 | 10 | 20 | 30 |
| Salvage value | 40 | -100 | 0 | 1000 |
| Annual savings | 60 | 400 | 100 | 400 |

are not a problem since each one will be compared to Doing Nothing. The equivalent present values may be computed for the corporate i*, for example 20%. For project A (automate packaging line), the cash flow diagram (expressed in \$1000s) is shown in Figure 4.4, and the equivalent present worth is found to be \$4500:

$$PW_A = -\$200 + \$60\left(\frac{P}{A}\right) + \$40\left(\frac{P}{F}\right)$$

$$= -\$200 + \$60\,(2.9906) + \$40\,(.4019)$$

$$= -\$4.5 \text{ (in \$1000s)}$$

Similarly, the cash flow diagrams for projects B, C, and D are shown in Figures 4.5, 4.6, and 4.7, while their corresponding equivalent present worths (in \$1000s) are found to be \$161, \$187, and \$396 respectively:

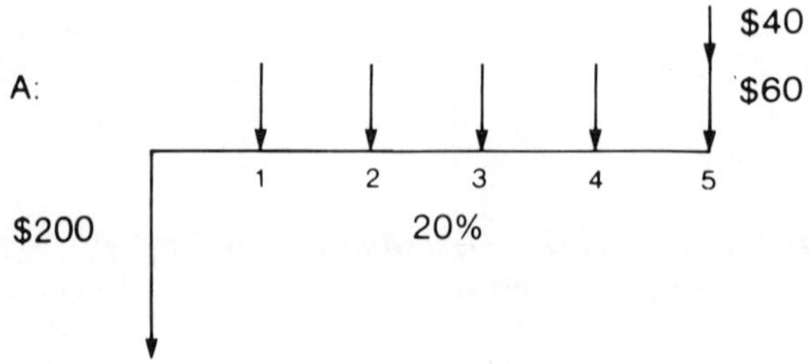


**Figure 4.4 Example 2: Automate packaging line (versus Do Nothing)**

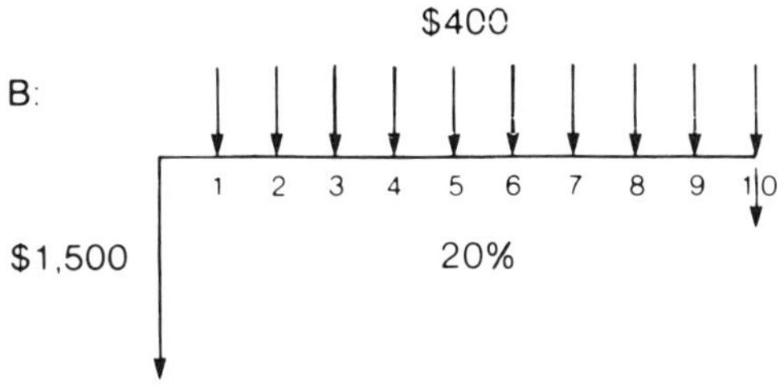


Figure 4.5 Example 2: Install heat recovery equipment

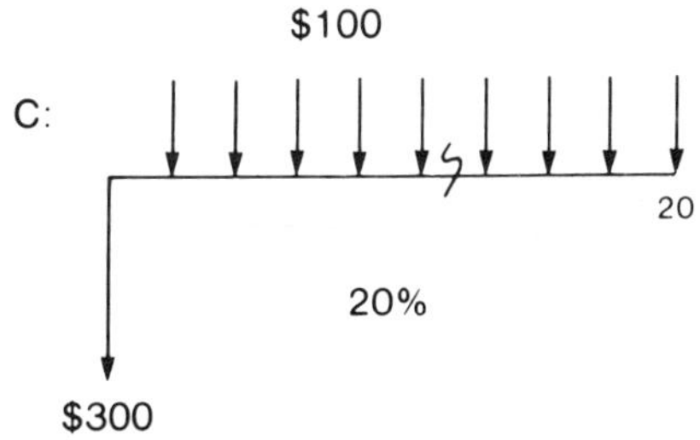


Figure 4.6 Example 2: Computerize inventory

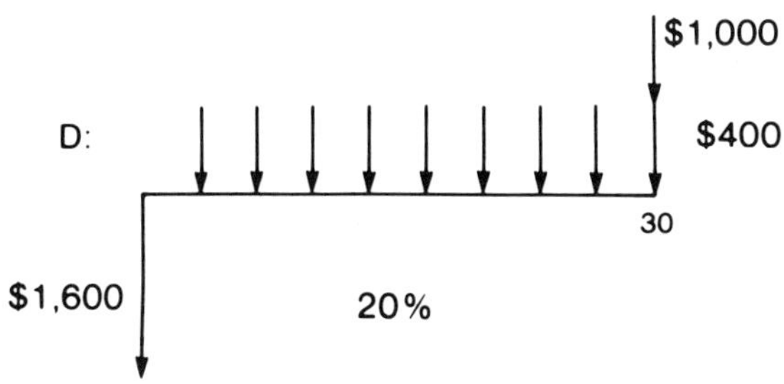


Figure 4.7 Example 2: Enlarge warehouse

$$PW_B = -\$1500 + \$400\left(\frac{P}{A}\right) - \$100\left(\frac{P}{F}\right)$$

$$= -\$1500 + \$400\,(4.1925) - \$100\,(.1615)$$

$$= -\$1500 + \$1677 - 16 = \$161$$

$$PW_C = -\$300 + \$100\left(\frac{P}{A}\right) = -\$300 + \$100\,(4.8696) = \$187$$

$$PW_D = -\$1600 + \$400\left(\frac{P}{A}\right) + \$1000\left(\frac{P}{F}\right)$$

$$= -\$1600 + \$400\,(4.9879) + \$1000\,(.0042)$$

$$= -\$1600 + \$1992 + 4 = \$396$$

The economically acceptable alternatives are all those with positive present worths: projects B, C, and D.

Several observations are in order. Notice that the net salvage value of option B is negative, indicating that an outward cash flow is required to remove the equipment. Also notice that at this relatively high i*, the salvage value in the distant future contributes little to the present worth. In fact, the estimate for option D's salvage value in 30 years could be in error by a factor of 10 with minor impact on the present worth. Finally we notice that the present worth method provides limited quantitative information to the decision maker regarding the extent to which each option exceeds the 20% minimum i*. For example, the present worths of projects B and C ($161 and $187) are of the same order, while the magnitudes of investment required ($1500 and $300) differ significantly. Although both projects exceed the i* cutoff, project B is less attractive than C, a fact that is not immediately obvious from the results of the present worth method.

### Annual Worth Method

The AW method is similar in concept to the PW method except that the differing alternatives are made comparable by converting their cash flows into equivalent uniform *annual* amounts. Again, the conversion is based on a minimum attractive rate i*, and the equivalent amounts are then compared.

For mutually exclusive alternatives, the best economic choice is the one with the greatest equivalent annual worth. For independent alternatives, all alternatives with positive annual worths are acceptable when judged against i*.

*Example 3: Solar System* To illustrate the annual cost method, we return to the solar system example, recalling that the annual heating cost without solar is $1000 and that i* is 10%. Summary data on the three alternative systems is given in Table 4.1.

The cash flow diagrams for the three solar systems as shown in Figures 4.1, 4.2, and 4.3 can be converted to their equivalent uniform annual worths:

$$AW_A = -\$2000\left(\frac{A}{P}\right) + \$300 = -\$2000\,(.1175) + \$300 = \$65$$

$$AW_B = -\$4000\left(\frac{A}{P}\right) + \$500 = -\$4000\,(.1175) + \$500 = \$30$$

$$AW_C = -\$10000\left(\frac{A}{P}\right) + \$900 + \$1000\left(\frac{A}{F}\right)$$

$$= (-10000)\,(.1175) + \$900 + \$1000\,(.0175) = -\$258$$

The solar options may now be ranked on the basis of their equivalent annual worths:

| *System* | *AW, $* |
|---|---|
| A | 65 |
| B | 30 |
| O | 0 |
| C | -258 |

The economic choice for these mutually exclusive alternatives is system A. Notice that the ranking is identical to that of the PW method. Again, the large system C is considerably worse than installing no solar system.

*Example 4: TransAmerican Industries* For the TransAmerican Industries example, the conversion of the differing cash flows to their equivalent annual values will give these results:

| *Project* | *AW, $1000s* |
|---|---|
| A Packaging line | -2 |
| B Heat recovery equipment | 38 |
| C Inventory system | 38 |
| D Warehouse | 79 |

Since the projects are independent, we may select any or all. The economically acceptable alternatives are those with positive annual worths: B, C, and D. Notice that the resulting annual worths tell us whether or not each project satisfies the i* cutoff, but provide little additional information. While the annual worths of projects B and C are identical ($38,000), their first costs and lives are very different, and it is difficult to judge their relative merits based on their annual worths.

## Background: Benefit/Cost and Return-on-Investment Methods

The PW and AW methods have much in common. Both are relatively easy to apply. Both accomplish pairing automatically, whether for only 2 alternatives or 200. Therefore, mutually exclusive alternatives are readily ranked, regardless of the number of alternatives. Both methods share the disadvantage of yielding results in dimensional dollars rather than nondimensional form. Consequently, neither is particularly helpful in judging the strength of an investment other than by its ability to satisfy the i* cutoff.

In contrast, the B/C and ROI methods are based on nondimensional scaling parameters, namely the B/C ratio and an interest rate (also a ratio). These ratios seem inherently easier for the human intellect to grasp and assimilate, especially when dealing with projects that differ greatly in magnitude. For example, saying that a project yields 25% usually conveys more information to the decision maker than stating the project's annual worth in absolute dollars.

All four methods–PW, AW, B/C, and ROI–are equally applicable to independent alternatives, where the proper comparison is always with Doing Nothing. It is with mutually exclusive alternatives that the nondimensional B/C and ROI methods present special problems. Unlike PW and AW, the pairing of dependent alternatives is not accomplished automatically with either the B/C or ROI method.

To expedite the required pairing, the alternatives are usually first ordered according to the magnitude of their initial investments. Then the incremental return or incremental B/C ratio for each stepup in investment is computed. If there are more than a handful of alternatives, the procedure is cumbersome when accomplished manually.

As a result of their strengths and weaknesses, the PW and AW methods lend themselves to the analysis of dependent alternatives while the ROI and B/C methods are more adaptable to independent alternatives.

## Benefit/Cost Method

With the PW and AW methods, all cash flows are merged algebraically to obtain their equivalent present or annual value at a given i*. With the B/C method, the benefit and cost streams are kept separate in computing their equivalent values. Then the equivalent benefits (expressed either in present or annual worths) are compared to the equivalent costs via the ratio of benefits to costs. If the equivalent benefits offered by the alternative outweigh the equivalent costs, then the B/C ratio is greater than 1 and the alternative is economically acceptable. If benefits fall short of costs, then the B/C ratio is less than 1 and the alternative fails. *The B/C standard is simply 1, that is, benefits must equal or exceed costs.*

To use the B/C method, one first must define benefits and costs. Several conventions are in use today, depending on the perspective of the user. The federal agencies usually define benefits as cash flows to or from the general public, while costs are government-incurred cash flows in the case of federally funded projects, or industry/utility cash flows in the case of regulating standards such as air quality or safety. Under this definition, the loss of a farmland to a flood control dam would constitute a negative benefit while the sale of sulfur recovered from stack emissions would be a negative cost.

Businesses often associate costs with the one-time cash flows of a capital investment (purchase or sale), while the benefits are taken as the annual savings and expenses. With this convention, the salvage value of an investment is a negative cost while maintenance is a negative benefit.

Regardless of one's choice of conventions, the B/C analysis always ranks given alternatives in the same order. Naturally the value of the B/C ratio will depend on the convention one adopts, although the B/C ranking does not.

*Example 5: TransAmerican Industries* To illustrate the B/C method, we recall the TransAmerican Industries example involving four independent projects, as shown in Table 4.2.

We will define costs as the cash flows associated with the one-time investment, while benefits are the cash flows associated with the ongoing operation. The benefits and costs each in turn will be converted to a comparable basis using equivalent annual values at an i* of 20%. The cash flow diagrams corresponding to the four alternatives are shown in Figures 4.4, 4.5, 4.6, and 4.7.

For project A (automate packaging line), the equivalent annual benefits and costs at an i* of 20% are found as follows:

$$B = \$60$$

$$C = \$200 \left(\frac{A}{P}\right) - \$40 \left(\frac{A}{F}\right) = \$300\,(.3344) - \$40\,(.1344) = \$62$$

$$\therefore \left(\frac{B}{C}\right)_A = \frac{\$60}{\$62} = .97 < 1 \qquad \text{unacceptable}$$

The B/C ratios for projects B, C, and D follow in like manner: For project B:

$$B = \$400$$

$$C = \$1500 \left(\frac{A}{P}\right) + \$100 \left(\frac{A}{F}\right) = \$1500\,(.2385) + \$100\,(.0385) = \$362$$

$$\therefore \left(\frac{B}{C}\right)_B = \frac{\$400}{\$362} = 1.10 > 1 \qquad \text{acceptable}$$

For project C:

$$B = \$100$$

$$C = \$300\left(\frac{A}{P}\right) = \$300\,(.2045) = \$61$$

$$\therefore \left(\frac{B}{C}\right)_C = \frac{\$100}{\$61} = 1.64 > 1 \qquad \text{acceptable}$$

For project D:

$$B = \$400$$

$$C = \$1600\left(\frac{A}{P}\right) - \$1000\left(\frac{A}{F}\right) = \$1600\,(.2009) - \$1000\,(.0009) = \$321$$

$$\therefore \left(\frac{B}{C}\right)_D = \frac{\$400}{\$321} = 1.25 > 1 \qquad \text{acceptable}$$

Notice that the B/C method yields the same conclusion as the PW and AW methods, namely, that project A is unacceptable; projects B, C, and D are acceptable. However, the nondimensional B/C ratio provides additional information to the decision maker for judging projects B, C, and D. Project C shows up strongly (B/C = 1.64), while project B (B/C = 1.10) may deserve a closer examination of the cash flow estimates.

The TransAmerican example involves independent alternatives. Consequently each project is compared with Do Nothing rather than against each other. The Do Nothing comparison is reflected in the above cash flow diagrams and resulting B/C ratios. The difficulty with the nondimensional B/C method (and the ROI method also) arises when the projects are mutually exclusive and must be compared each with the other. To illustrate the care required when dealing with mutually exclusive alternatives, we will reexamine the solar energy example.

*Example 6: Solar System* Information on the three systems of differing size is summarized in Table 4.1. No more than one system will be chosen.

We begin by computing the B/C ratio on the total investment for each solar system, that is, compared to Do Nothing (an incomplete comparison since the systems compete with each other). Again we will define costs as relating to capital investment and benefits as relating to annual benefits and expenditures. Figures 4.1, 4.2, and 4.3 show the cash flow diagrams. Utilizing equivalent annual costs and benefits at an i* of 10%, the B/C ratio may be computed. For the small system A, the B/C ratio is 1.28:

$$B = \$300$$

$$C = \$2000\left(\frac{A}{P}\right) = \$2000\,(.1175) = \$235$$

$$\therefore \left(\frac{B}{C}\right)_A = \frac{\$300}{\$235} = 1.28$$

Similarly, the B/C ratios for systems B and C are found to be 1.06 and .78:

$$B = \$500$$

$$C = \$4000\left(\frac{A}{P}\right) = \$4000\,(.1175) = \$470$$

$$\therefore \left(\frac{B}{C}\right)_B = \frac{\$500}{\$470} = 1.06$$

System C:

$$B = \$900$$

$$C = \$10{,}000\left(\frac{A}{P}\right) - \$1000\left(\frac{A}{F}\right)$$

$$= \$10{,}000\,(.1175) - \$1000\,(.0175) = \$1158$$

$$\therefore \left(\frac{B}{C}\right)_C = \frac{\$900}{\$1158} = .78$$

At first glance, it may appear that either A or B is acceptable since the benefits outweigh the costs in each. Based on total investment (that is, versus Do Nothing), they are both acceptable. But is B worth the extra investment *when compared to A*?

If the alternatives were independent, then both would be acceptable. Since only one may be selected, then they must be paired against each other. Unlike the AW and PW methods, the B/C and ROI methods do not accomplish this pairing automatically. Instead we must explicitly consider the *incremental* benefits and *incremental* costs between A and B. The difference between A and B is reflected by their difference in cash flows, as shown in Figure 4.8. The economic question now posed is: Should we increase our investment by \$2000 in exchange for \$200 greater annual savings? The *incremental* B/C ratio must be determined:

$$B = \$200$$

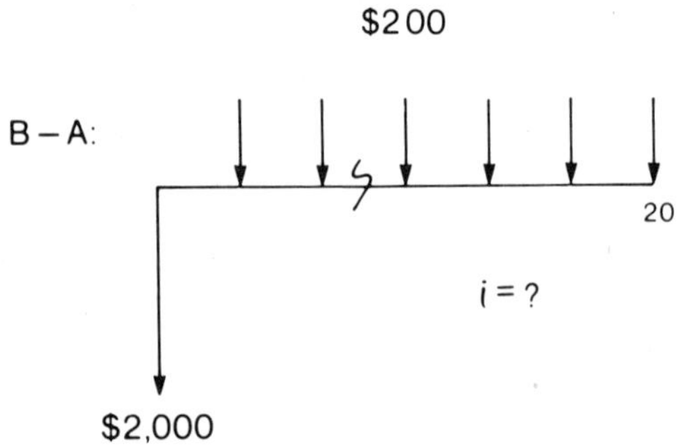


**Figure 4.8 Solar system B versus A for B/C and ROI analysis**

$$C = \$2000\left(\frac{A}{P}\right) = \$2000\,(.1175) = \$235$$

$$\therefore \left(\frac{B}{C}\right)_{B-A} = \frac{\$200}{\$235} = .85 < 1$$

The extra $2000 investment is clearly not acceptable. The economic choice is therefore system A–a conclusion agreeing with the present and annual worth analyses.

You may be tempted to conclude that one should simply select the highest B/C ratio, in this case system A. This is a common error. The selection criterion is not to maximize the B/C ratio but rather to maximize the investment that results in a B/C greater than 1. In other words, if the extra investment yields extra benefits exceeding the extra costs, then the extra investment is economically sound.

If one must compare more than handful of mutually exclusive alternatives without the aid of a computer, then the B/C method becomes cumbersome. One way to expedite the explicit comparison of all possible pairs is to order the alternatives by increasing investment, then examine each incremental investment in turn. If alternative B fails when compared to A, then discard it and go on to C, again comparing with A. If C passes then compare D with C, and so on.

The B/C method is particularly popular in the public sector, probably since it was institutionalized in the 1930s by congressional legislation that authorized flood control projects. Unfortunately some legislators fail to understand the importance of incremental analysis, choosing instead to favor the highest B/C ratio.

The application of engineering economics in the public sector is fraught with difficulties, whether one uses B/C or another method. The minimum attractive return rate i* (often called the *discount rate* for public projects) is frequently set at a relatively low value, a value comparable to the yield on long-term government securities. Critics argue that a higher value reflecting the taxpayers' i* is more appropriate. Since the costs of a project tend to be concentrated at the front end (for example, construction of a flood control and recreation dam) while the benefits occur in the future, a low i* tends to discount the benefits less. Consequently, the project appears more attractive than it would at a higher i*.

The most challenging difficulty with the B/C analysis of a public project is in quantifying the benefits (that is, the cash flows to and from the public). Establishing one's viewpoint, correctly identifying the appropriate benefits, and then attaching a dollar value to each benefit requires human judgments which are in part based on opinions. Differences in opinions lead to sharp conflicts ultimately resolved in the political arena.

## Return-on-Investment Method

The ROI method, like the B/C method, yields a nondimensional ratio that may be compared to standard criteria. With the ROI method, the rate of return is computed and compared with the minimum acceptable rate i*. Investments not equaling or exceeding i* are discarded. As with the B/C method, an explicit incremental analysis is required for mutually exclusive alternatives.

To illustrate the ROI method, we again use the TransAmerican Industries example (less project D) involving independent alternatives, and the solar system example with its dependent alternatives.

*Example 7: TransAmerican Industries* Table 4.2 summarizes the TransAmerican projects. Again we compare these independent alternatives with Do Nothing. Figures 4.4, 4.5, and 4.6 show the cash flow diagrams (relative to Do Nothing). With the ROI method, we compute the interest rate i which satisfies the equation for the cash flow diagram and compare to the i* of 20%. The solution is trial and error. For project A, the overall return rate is found to fall between 15% and 20%:

$$\$200 = \$60 \left(\frac{P}{A}\right) + \$40 \left(\frac{P}{F}\right)$$

Try i = 20%:

$$\$200 \stackrel{?}{=} \$60\,(2.9906) + \$40\,(.4019)$$

$$\$200 > \$196$$

Try i = 15%:

$$\$200 \overset{?}{=} \$60\,(3.3522) + \$40\,(.4972)$$

$$\$200 < \$221$$

Interpolating, the equivalent interest rate or ROI for project A is about 19%. The ROI fails to meet the required 20% cutoff.

Similarly, project B is found to return about 23%:

$$\$1500 = \$400\left(\frac{P}{A}\right) - \$100\left(\frac{P}{F}\right)$$

Try i = 20%:

$$\$1500 \overset{?}{=} \$400\,(4.1925) - \$100\,(.1615)$$

$$\$1500 < \$1661$$

Try i = 25%:

$$\$1500 \overset{?}{=} \$400\,(3.5705) - \$100\,(.1074)$$

$$\$1500 > \$1417$$

Interpolating the tables, the ROI for project B is about 23% thereby satisfying the 20% cutoff. For project C:

$$\$300 = \$100\left(\frac{P}{A}\right)$$

In this case, we can solve for i explicitly:

$$\frac{P}{A} = \frac{\$300}{\$100} = 3.0$$

From the tables

$$\frac{P}{A} = 3.3158 \qquad \text{for 30\% and N = 20}$$

$$\frac{P}{A} = 2.4970 \qquad \text{for 40\% and N = 20}$$

Therefore, project C offers a return of about 34%, well above the 20% i*.

*Example 8: Solar System* We begin Example 8 by computing the ROI on the total investment for each solar system, that is, compared to Do Nothing. Again, this comparison is inadequate since the systems compete with each other as well as Do Nothing. See Table 4.1 and the cash flow diagrams (Figures 4.1, 4.2, and 4.3.) The cash flow equation for system A may be solved for i:

$$\$2000 = \$300 \left(\frac{P}{A}\right)$$

$$\therefore \frac{P}{A} = 6.67$$

Interpolating from the tables:

$$\frac{P}{A} = 6.26 \qquad \text{for } i = 15\% \text{ and } N = 20$$

$$\frac{P}{A} = 7.47 \qquad \text{for } i = 12\% \text{ and } N = 20$$

The ROI for system A on the total investment is about 14%, exceeding the 10% minimum. Therefore system A is acceptable. For system B:

$$\$4000 = \$500 \left(\frac{P}{A}\right)$$

$$\therefore \frac{P}{A} = 8.00$$

Interpolating from the tables:

$$\frac{P}{A} = 7.4695 \qquad \text{for } i = 12\% \text{ and } N = 20$$

$$\frac{P}{A} = 8.5136 \qquad \text{for } i = 10\% \text{ and } N = 20$$

Therefore the overall return on system B (total investment) is about 11%. It is too early to declare B as acceptable since we have not yet compared B with A. To determine if the incremental investment required for B over A is justified, we must examine the difference between A and B. Figure 4.8 illustrates the cash flow difference between A and B. It shows that the question is: Should we invest \$2000 more now in return for \$200 greater annual savings?

$$\frac{P}{A} = \frac{\$2000}{\$200} = 10.0$$

From the interest tables, we find that the incremental ROI corresponding to a P/A of 10.0 is less than 8%—below the acceptable limit. System B is rejected.

The next more expensive alternative is system C. It should now be compared to the last acceptable alternative, that is, system A. Figure 4.9 shows the cash flow differences between C and A. The solution for i is trial and error:

$$\$8000 = \$600\left(\frac{P}{A}\right) + \$1000\left(\frac{P}{F}\right)$$

Try 5%:

$$\$8000 \stackrel{?}{=} \$600\,(12.4622) + \$1000\,(.3769)$$

$$\$8000 > \$7854$$

Try 3%:

$$\$8000 \stackrel{?}{=} \$600\,(14.8775) + \$1000\,(.5537)$$

$$\$8000 > \$9480$$

The incremental ROI on system C relative to A is slightly less than 5%: It is rejected.

Notice that the ROI method yields a nondimensional number (namely, the return rate), that is more supportive of the intellectual thought processes

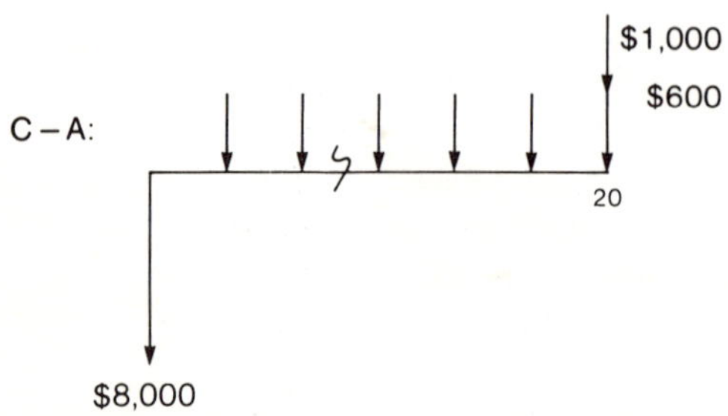


**Figure 4.9 Solar system C versus A for B/C and ROI analysis**

required for decisions. The ROI is easier to grasp and provides more information than either the AW or PW methods. However, it is less convenient to use with dependent alternatives because it requires explicit treatment of incremental investment.

Recent surveys report that the ROI method is utilized by more U.S. corporations than any other capital budgeting method, especially for judging independent alternatives.

## Pseudomethod: Years-to-Payback

If you are dismayed by the complexities of engineering economy, if you want a simple answer to your capital budgeting problems, if you were hoping to understand completely economic analysis after a few minutes of easy reading, then years-to-payback method is for you.

The payback method is simple, easy to understand and to communicate. It will amaze your friends and impress your business associates. They may actually believe that you are knowledgeable in economic analysis. But alas, the years-to-payback method is superficial in its approach, oversimplified in its results, and theoretically unsound. In other words, it will often lead to bad decisions.

But the payback method is popular. It is popular because it is so simple, quick, and easy to communicate. For those who must slay dragons and fight fires daily and have little time for economic complexities, the temptation to utilize years-to-payback may be irresistible. Because of its popularity, the payback method deserves our attention.

The number of years-to-payback is simply the years required for the positive cash flows to equal the initial cost.

Consider three, 6-year investments having the anticipated cash flows illustrated in Table 4.3. Question: What is the years-to-payback for each of these projects? The answer in all three cases is 3 years. After 3 years, the sum of the positive cash flows equals the negative cash flow. All three proposals offer identical payback periods. But are they all equally attractive? Obviously, they are not. Project A is clearly superior to C since it yields benefits the full 6 years. And project B is superior to A, unless the interest rate is extraordinarily high. Only during the first year does A exhibit an advantage over B. And yet the payback method rates A, B, and C equally!

Do you see the serious deficiences of the payback method? They are twofold:

Payback normally does not discount future cash flows. Since i is zero, all cash flows are weighted equally, regardless of their timing.

**Table 4.3 Cash flow table illustrating the years-to-payback method**

| | Alternative | | |
|---|---|---|---|
| End-of-year | A | B | C |
| 0 | -30 | -30 | -30 |
| 1 | 10 | 8 | 10 |
| 2 | 10 | 10 | 10 |
| 3 | 10 | 12 | 10 |
| 4 | 10 | 14 | 0 |
| 5 | 10 | 16 | 0 |
| 6 | 10 | 18 | 0 |

Payback truncates the cash flow pattern at the end of the payback period, ignoring differences after truncation.

Some users modify the payback method by introducing an i* for discounting future cash flows. While this modification eliminates the first deficiency, it also sacrifices the great simplicity of the method, making it much less attractive to the simplistic-minded decision maker. And the modification does nothing to rectify the second deficiency, which is inherent to the payback method. Since years-to-payback ignores cash flow differences occurring after payback, it unjustly penalizes longer-term projects. Consequently it may lead to erroneous decisions.

Some companies employ the payback method in conjunction with another method such as ROI. Their justification is that it helps ensure against the risk of a project's early unexpected demise. And it is so simple!

The four methods–PW, AW, B/C, and ROI–will lead to identical economic conclusions when applied correctly. The PW and AW methods are the easiest to use, at least without the assistance of a programmed computer, but they yield essentially yes/no information. The dimensionless methods of B/C and ROI provide more information but require explicit incremental analysis for dependent alternatives. The choice among the four methods is a matter of personal preference or corporate policy.

Whatever your choice, the method should be applied over the entire life cycle of the project, and alternative projects should always be compared over equal time periods.

# 5

# Taxes and Accounting

*The Congress shall have powers to lay and collect taxes . . . .*
U.S., *Constitution,* sec. 8.

## Our Partnership with Government

From the earliest moments of recorded civilization, human beings have organized themselves into societies characterized by specialized labor, common language and exchange, and a governing structure. Government provides stability and order. Through order, we may live more productively, with less tension and anxiety. Government provides essential services together with a mechanism for balancing the rights of its citizens.

Our country's forefathers envisioned a unique form of government, one "of the people, by the people and for the people," designed to "established Justice, insure domestic Tranquility, provide for a common defense, promote the general Welfare, and secure the Blessings of Liberty to ourselves and our Prosterity."

The cost of that government is taxes—the compulsory payment in support of government. Income tax, payroll tax, property tax, sales tax, excise tax, taxes on gross receipts, on inventory, on tangible assets—these taxes and others are all mandatory, for the support of our government. In a very real sense, we labor *in partnership with government.* It shares our profits, it shares our risks, it shares our assets and our losses.

When the private sector of our economy is healthy, when profits are high and unemployment low, then government shares the benefits through increased tax revenues. In contrast, when individual and corporate profits stagnate and unemployment climbs, then the government's share of the partnership suffers also. Unfortunately, unemployment compensation, welfare, and other social service demands on government increase during difficult economic times.

Corporate tax cuts initiated by the President and Congress in the early 1960s and in 1980 were in recognition of this essential partnership between business and government. The tax cuts strengthened the profit motivation of the private sector, increased productivity, and resulted in higher profits and tax revenues. Both partners benefited.

## Taxes: An Important Cash Flow

Engineering economy focuses on *cash flow.* We identify the comings and goings of money for a given investment alternative and adjust for timing differences via an appropriate interest rate. Identification of the cash flows requires a clear definition of our economic system. In other words, we must clearly establish our perspective for the economic analysis.

The usual perspective for private enterprise excludes government from the system. We usually concentrate on maximizing our own investment return, knowing that our government partner will receive its fair share. From this perspective, taxes are viewed as cash flows leaving the system, and the economic analysis must be an *after-tax* analysis.

The impact of taxes on the viability of a proposed investment may be crucial. Investments that appear comparable before taxes may be very different when viewed after taxes. Tax rates may be high. The federal income tax rate for individuals increases progressively with taxable income from 14% to a high of 70%. State and local income taxes may add another 4% to 8%. The maximum corporate income tax rate currently increases from 17% of the first $25,000 to 46% of all taxable income over $100,000.

Obviously the amount of one's income tax obligation is a function of one's profitability. More precisely, it is a function of *taxable income.* But taxable income is an accounting number. *Taxable income is determined from the accounts of the taxpayer.* While taxes themselves are important cash flows, taxable income is an accounting number, not cash flow per se.

Our primary concern is cash flow. Nevertheless, we must digress temporarily into principles of accounting as they determine the important cash flow of taxes. One's choice of accounting procedures affects the after-tax return on an investment by affecting taxes.

## Accounting and Engineering Economy

The engineering economist looks into the future. Proposed investments are compared on the basis of their projected cash flows. In contrast, the accountant is a historian who records past events. The primary objective of the accountant is to represent fairly the financial position and results of operation to management, owners, and investors. However, the representation is based on past events and may be distorted by a rapidly changing environment. For example, land

and inventory assets are normally accounted for at acquisition cost. During periods of severe inflation, the accounts may grossly undervalue these assets. If cost of goods is increasing rapidly, the accountant's choice for evaluating inventory, last in/first out or first in/first out (LIFO or FIFO), will influence the net profits shown on the firm's accounts.

The engineering economist deals with cash flows as they occur. Although the accountant's books ultimately relate to cash flows, they do not portray accurately their timings. For example, the cash flow needed to construct a new plant occurs early in the project, when the actual expenditure occurs. But the accountant will distribute this first cost over many years in an attempt to represent fairly the results of plant operation.

Since the viewpoints of the economist and accountant are quite different, one must be selective in utilizing accounting numbers as cash flow. However, income taxes (a very real cash flow) can only be determined from the accounts, since income tax is computed from taxable income, an accounting number.

## The Statement of Income

In order to grasp how accounting influences tax flow, we must have at least a rudimentary knowledge of the accountant's *statement of income.* The income statement portrays the results of operation over a particular time period, for example, January 1 through December 31st. The basic structure of a typical income statement is shown in Table 5.1.

**Table 5.1 Example of an income statement**

| Statement of Income |
|---|
| Gross income: |
| Sales |
| Other |
| - Costs and expenses: |
| Costs of products sold |
| Marketing and administration |
| Interest |
| = Income before taxes |
| - Operating taxes: |
| Federal income |
| State/local income |
| Property/payroll/other |
| = Net Income |

A taxpayer's taxable income is directly related to the "Income Before Taxes" on the income statement. In the case of an individual taxpayer, certain exemptions and deductions are subtracted from gross income to obtain taxable income. For the corporate taxpayer, the situation is more complex but the principle is unchanged: certain expenses incurred in earning income may be subtracted from gross income before computing taxes. Some of these expenses are cash flows that have occurred during the time period of the statement. Others may be related to cash flows that occurred in previous periods or are anticipated to occur in future periods (for example, accounts payable). A major expenditure for many corporations, namely *dividends* does not appear as an expense item on the income statement and therefore is not tax deductible. In the U.S., dividends have always been considered as distribution of net income to the owners rather than as a business expense.

It should be apparent that rather detailed accounting standards are required for defining allowable expenses. These standards are important in that they influence (1) the financial image presented by the accounts, and (2) income taxes.

### Securities and Exchange Commission and Internal Revenue Service

Two federal agencies—the Securities and Exchange Commission (SEC) and the Internal Revenue Service (IRS)—are keenly interested in the accounting standards utilized for determining the statement of income.

A principal function of SEC is to ensure that all relevant facts pertaining to an investment security are made available to prospective investors. The SEC concern is that the public accounts fairly represent the financial position "in accordance with generally accepted standards." Until recently, these accounting standards were formulated by the Accounting Principles Board of the American Institute of Certified Public Accountants (AICPA). More recently, the new Financial Accounting Standard Board has become the arbiter.

The IRS represents the federal government's partnership in our financial endeavors. As our partner, the IRS is interested in our profitability and its determination. The standards legislated by Congress and administered by the IRS are complex, as suggested by the comprehensive IRS listing of taxpayer documents, *Catalog and Index to Taxpayer Information Publications,* IRS Publication 900. Two booklets are particularly valuable to many taxpayers and are available free from IRS: *Your Federal Income Tax (for Individuals),* IRS Publication 17, and *Tax Guide For Small Businesses,* IRS Publication 334.

The Internal Revenue Codes and interpretive taxpayer documents describe in detail the constraints placed on accounting for tax purposes. Usually the tax accountant attempts to minimize the taxpayer's obligation within IRS constraints. In contrast, the management accountant attempts to provide accurate financial information to management and owners for control of the enterprise. Frequently management feels that the tax accounts inadequately portray the financial position of the enterprise for purposes of administrative control. As a result, most U.S. corporations now maintain two legal sets of books, one for financial control and public disclosure and one for tax purposes. The tax accounts are important to the engineering economist in that they determine the tax cash flows for an investment.

# 6

# Income Tax Considerations

## Accounting Impact on Investment Economics

While the accountant is a financial historian whose viewpoint of cash flow differs from that of the engineering economist, the accountant is not just an impartial observer. The accounting standards adopted may have a crucial impact on the economics of an investment alternative.

The reason is taxes. Income taxes are computed on the basis of taxable income, and taxable income is an accountant's number. Consequently those charged with the responsibility of capital budgeting must appreciate the impact of accounting on taxes and thereby on investment economics. Two tax effects are particularly important:

Writeoff of productive capital assets (termed *depreciation* and *depletion*), and

Tax credits

## Writeoff of Capital Assets: Depreciation

In the wilderness of words that comprises the English language stand those majestic sequoias distinguished by their multiple meanings. *Depreciation* is one such word. Physical degradation and deterioration, loss in market value, decline in usefulness, replacement trade-in cost—these conflicting meanings are all in widespread use. But the accountant defines depreciation in yet another way, as the *writeoff of an asset's costs over its life, in a systematic and rational manner.*

A fundamental principle of taxation is that certain expenses incurred in the production of income are tax-deductible, that is, excluded from taxable income. The day-to-day operating expenses of doing business are deductible as they occur: labor, materials, maintenance, utilities, taxes, insurance, and interest charges. But certain expenditures are investments that serve for many years, and improve (rather than maintain) the physical facilities. Examples of these *capital investments* include buildings, machinery, trucks, office equipment, and patents. Rather than expense these capital expenditures as they occur, the accountant charges the expenditure off over the asset's life in some systematic and rational manner. This capital writeoff is called depreciation and appears as a tax-deductible expense on the income statement. For natural resources such as mineral deposits and timber, the writeoff is called *depletion.*

## Writeoff Methods

Since depreciation is tax-deductible, Congress has established accounting standards for its determination. These standards include guidelines and constraints on the depreciation base, salvage value, writeoff life, and writeoff method.

The *depreciation base* is usually the cost of the property. If you materially improve the property, the additional costs are added to the base, while casualty losses reduce the base. Land is never depreciated.

The *salvage value* is the amount that you estimate will be realized upon disposal of the asset. It is estimated at the time of acquisition and is usually not subsequently adjusted due to price level changes. Salvage, when reduced by the cost of removal and disposal, is called *net salvage.* Either salvage or net salvage may be used in determining depreciation so long as one is consistent. However, a net salvage value less than zero is not permitted for writeoff purposes. No asset may be depreciated below a reasonable salvage value under any method.

*Writeoff life* is the time period over which the asset may reasonably be expected to be useful to the taxpayer for the production of income. The writeoff life may be determined from the particular operating conditions and experience of the taxpayer or from the general experience in the industry, if the taxpayer's experience is inadequate.

For property acquired after 1970, the taxpayer may elect to depreciate in accordance with guideline lives established as part of the Class Life Asset Depreciation Range (CLADR) system. The CLADR system is designed to minimize disputes between the taxpayer and IRS. Under the CLADR system, salvage value is not considered in computing the annual writeoff, and the taxpayer must abide by the guideline lives. Sample guidelines are shown in Table 6.1. The CLADR system is described in detail in IRS Publication 534, *Tax Information on Depreciation.*

Three writeoff methods are widely used for computing depreciation:

Straight-line

Declining balance

Sum-of-the-years digits

**Table 6.1 Examples of guideline lives for the class life asset depreciation range (CLADR) system[a]**

| | Depreciation range, yr | | |
|---|---|---|---|
| Asset | Lower limit | Guideline | Upper limit |
| Automobiles | 2.5 | 3 | 3.5 |
| Tractors, over-the-road | 3 | 4 | 5 |
| Computers, peripheral equipment | 5 | 6 | 7 |
| Airplanes | 5 | 6 | 7 |
| Cattle, breeding/dairy | 5.5 | 7 | 8.5 |
| Agriculture machinery | 8 | 10 | 12 |
| Wholesale/retail trade | 8 | 10 | 12 |
| Office furniture | 8 | 10 | 12 |
| Personal/professional services | 8 | 10 | 12 |
| Manufacture of chemical products | 9 | 11 | 13 |
| Railroad cars | 12 | 15 | 18 |
| Roads, sidewalks, sewers | | 20 | |
| Farm buildings | 20 | 25 | 30 |

[a]Publication 534, *Tax Information on Depreciation*, U.S. Internal Revenue Service, Department of the Treasury.

The straight-line (SL) method is the simplest to use. Under SL depreciation, the property cost less salvage value* is written off in uniform annual amounts over the writeoff life:

$$\text{Annual writeoff (SL)} = \frac{\text{first cost} - \text{salvage}}{\text{life}} \tag{1}$$

Under the declining-balance method each year's writeoff is subtracted from the property's remaining balance before computing the next year's writeoff, so that the same depreciation rate is applied to a smaller or declining balance each year. The maximum depreciation rate is double that of the SL method. For certain used property and real estate, the maximum rate is 1½ or 1¼ the SL rate. As an example, for a life of 5 years, the double-rate-declining-balance (DRDB) rate is 40% or double the SL rate of 20%. Salvage value is not deducted

*Under the CLADR system, salvage value is zero.

from the cost in determining the declining balance writeoff, but property may not be depreciated below its estimated salvage value:

$$\text{Annual writeoff (DRDB)} = \frac{\text{remaining balance}}{\text{life}} \times 2.0 \tag{2}$$

With the sum-of-the-years-digits method (SOYD), a decreasing fraction of the depreciable amount is written off each year. The depreciable amount is the property cost less its salvage value. The denominator of the fraction is the sum of all the years in the life. For example with a 5-year life, the sum is *15* (1 + 2 + 3 + 4 + 5 = 15), while for the life of N years, the sum of the digits is N(N + 1)/2. The numerator is the years in reverse order so that the writeoff fractions for a 5-year asset would be:

| *Year* | *Writeoff fraction* |
|---|---|
| 1 | 5/15 |
| 2 | 4/15 |
| 3 | 3/15 |
| 4 | 2/15 |
| 5 | 1/15 |
| Total | 15/15 |

The annual writeoff is the writeoff fraction of the first cost less salvage:

$$\text{Annual writeoff (SOYD)} = (\text{first cost} - \text{salvage}) \times (\text{writeoff fraction}) \tag{3}$$

***Example 1: Straight-line, Double Rate Declining Balance, and Sum-of-the-Years-Digits*** To compare the three common writeoff methods, consider this simple example:

First cost = \$10,000

Writeoff life = 10 yr

Salvage value = 0

The SL writeoff is a uniform \$1000 for each of the 10 years:

$$\text{Annual writeoff} = \frac{\$10000 - 0}{10} = \$1000$$

For DRDB, the writeoffs for the first 3 years are:

$$\text{First year writeoff} = \frac{\$10000}{10} \times 2 = \$2000$$

$$\text{Second year writeoff} = \frac{\$8000}{10} \times 2 = \$1600$$

$$\text{Third year writeoff} = \frac{\$6400}{10} \times 2 = \$1280$$

While for SOYD, the corresponding writeoffs are:

$$\text{Sum} = N\left(\frac{N+1}{2}\right) = \left(\frac{10 \times 11}{2}\right) = 55$$

$$\text{First-year writeoff} = (\$10000 - 0)\left(\frac{10}{55}\right) = \$1818$$

$$\text{Second-year writeoff} = (\$10000 - 0)\left(\frac{9}{55}\right) = \$1636$$

$$\text{Third-year writeoff} = (\$10000 - 0)\left(\frac{8}{55}\right) = \$1455$$

The results of Example 1 are summarized graphically in Figure 6.1.

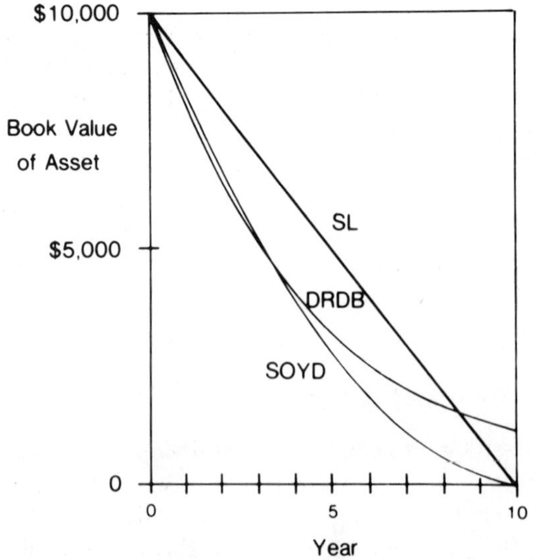


**Figure 6.1 Example 1: Comparison of three writeoff methods (zero salvage)**

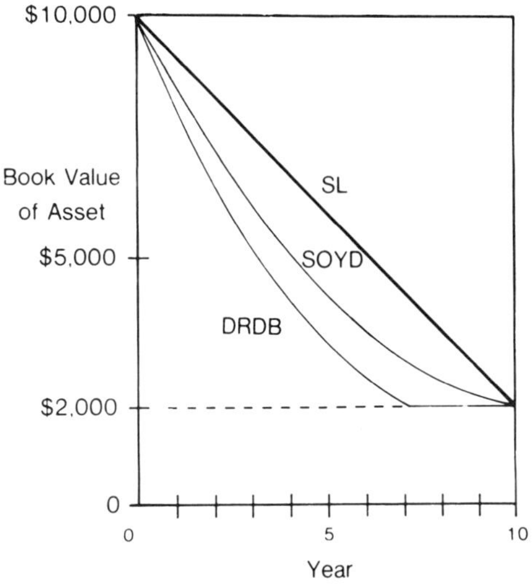


**Figure 6.2 Comparison of three writeoff methods with nonzero salvage**

If this example is modified by increasing the salvage value to $2000, the relative writeoffs are changed significantly as shown in Figure 6.2. The DRDB method now yields the most rapid writeoff.

The SL method of depreciation obviously expenses the asset uniformly over its life, while both the DRDB and SOYD methods result in more rapid, or accelerated, depreciation. Compared to SL, the accelerated writeoff methods result in larger tax deductions in the early years of the project and smaller tax deductions in the later years. As a result, the taxes for an income producing investment are deferred until later years. For those firms that use accelerated writeoff for tax accounting and SL writeoff for financial accounting, these deferred taxes will normally appear on the financial balance sheet of the firm as a liability.

## The Economic Advantage of Accelerated Writeoffs

Clearly one's choice of writeoff procedures affects the timing of the tax cash flows. For taxpayers whose tax bracket is constant or slowly changing from year to year, as is the case for most individuals and corporations, accelerated writeoff delays the timing of taxes without materially changing the amount and is therefore advantageous. The greater the taxpayer's time value of money, the more valuable is the delay.

To illustrate the advantage of accelerated writeoff to the taxpayer with a relatively constant tax rate, we will consider two examples:

Effect of writeoff method on after-tax return on investment, and

Effect of writeoff life

*Example 2: Effect of Writeoff Method* Let us assume that the \$10,000 investment in Example 1 will result in before-tax cash savings of \$3000/year:

| | |
|---|---|
| First cost | = \$10,000 |
| Salvage | = 0 |
| Life | = 10 yr |
| Before-tax savings | = \$3000/yr |
| Income tax rate | = 46% |

The before-tax economic analysis using the return-on-investment (ROI) method is straightforward. We first sketch the before-tax cash flow diagram, as shown in Figure 6.3. The A/P factor may be seen to be 0.30. Interpolating from the interest tables (N = 10), we find that the expected before-tax return on this proposed investment is *27.3%*.

Our partner, the government will share in the project savings through taxation. Consequently, the before-tax savings will be reduced by the amount of taxes. Each year's taxes may be computed from the savings less depreciation for that year. For example, with SL depreciation, the annual writeoff is \$1000/year. This accounting expense reduces the taxable portion of the before-tax savings to \$2000/year (\$3000 less \$1000). At a tax rate of 46%, the income tax is \$920 (\$2000 × .46), leaving an after-tax savings of \$2080 (\$3000 less \$920).

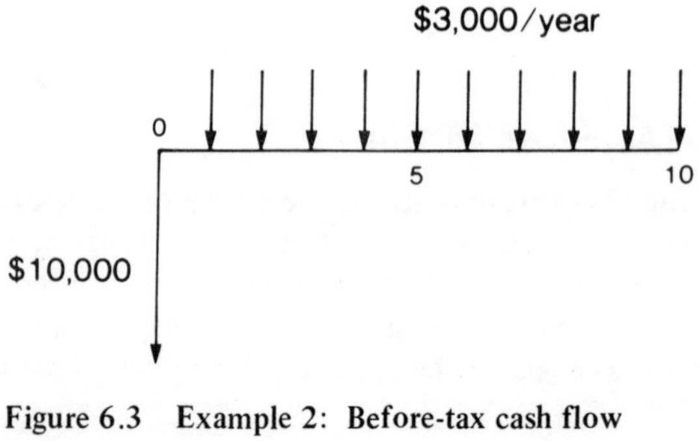


**Figure 6.3 Example 2: Before-tax cash flow**

Table 6.2 Writeoff and after-tax savings for SL, DRDB, and SOYD writeoff in Example 2

| Year | SL writeoff/savings, $ | DRDB writeoff/savings, $ | SOYD writeoff/savings, $ |
|---|---|---|---|
| 1 | 1000/2080 | 2000/2540 | 1818/2456 |
| 2 | 1000/2080 | 1600/2356 | 1636/2373 |
| 3 | 1000/2080 | 1280/2209 | 1455/2289 |
| 4 | 1000/2080 | 1024/2091 | 1273/2206 |
| 5 | 1000/2080 | 819/1997 | 1091/2122 |
| 6 | 1000/2080 | 655/1921 | 909/2038 |
| 7 | 1000/2080 | 524/1861 | 727/1954 |
| 8 | 1000/2080 | 419/1813 | 545/1871 |
| 9 | 1000/2080 | 336/1775 | 364/1787 |
| 10 | 1000/2080 | 268/1743 | 182/1704 |
| Total | 10,000/20,800 | 8925/20,306 | 10,000/20,800 |

The depreciation and after-tax savings for each of these writeoff methods are summarized in Table 6.2.

Notice that the *total* writeoff and *total* after-tax savings for SL depreciation and SOYD are identical. But they differ in timing. As an accelerated writeoff, SOYD realizes the cash savings earlier. Consequently, we will find that the return on investment for the proposed project is greater when SOYD accounting is utilized. Likewise, the DRDB method yields a higher return than SL, although the totals are somewhat different since salvage value is not useful in computing the DRDB writeoff.

To determine the investment return corresponding to each of the deprecia-tion methods, we may solve the after-tax cash flow diagrams. For example, the after-tax diagram for SL deprecation is shown in Figure 6.4.

Interpolating from the interest tables for an A/P factor of .208 corresponding to the figure, we find that the return is *16.1%*.

The solutions for the DRDB and SOYD cash flow diagrams are somewhat more difficult if accomplished manually, due to the nonuniform cash flows. The results are summarized below:

*Before-tax return on investment*

All 3 methods: 27.3%

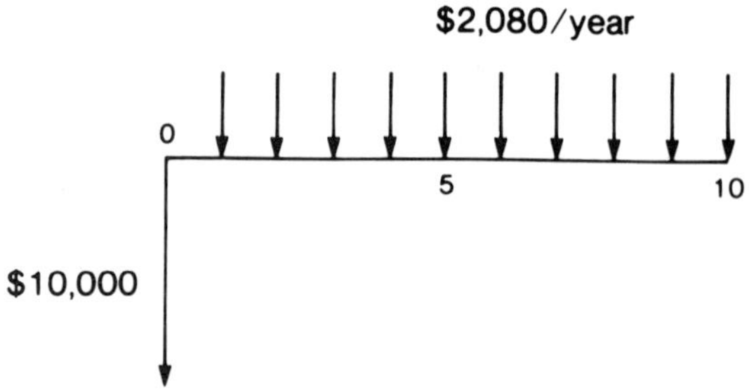


**Figure 6.4 Example 2: After-tax cash flow assuming SL depreciation**

*After-tax return on investment*

| | |
|---|---|
| SL: | 16.1% |
| DRDB: | 16.4% |
| SOYD: | 17.7% |

Notice that an identical investment yields significantly different after-tax returns for different accounting writeoff methods. In this example, the salvage was assumed for simplicity to be zero. In those situations where the salvage value is a significant fraction of the first cost, the acceleration effect of DRDB is strengthened relative to SL and SOYD, since the DRDB writeoff is not reduced by salvage.

*Example 3: Effect of Writeoff Life* Depreciation may be accelerated by the use of a shortened writeoff life as well as an accelerated writeoff method. By authorizing shortened guideline lives, Congress has encouraged capital investments in order to stimulate American productivity.

As an example of the impact of writeoff *life* on investment return, we will extend the previous example to several writeoff lives. For simplicity, SL depreciation will be utilized throughout the example:

*Proposed capital investment*

| | |
|---|---|
| First cost | = $10,000 |
| Net salvage | = 0 |
| Economic life | = 10 yr |
| Before-tax savings | = $3000/yr |
| Income tax rate | = 46% |

**Table 6.3 Cash flow and accounting table for Example 3 with 5-year writeoff and 10-year economic life**

| End-of year | Before-tax cash flow, $ | Writeoff | Taxable income, $ | Tax @ 46% | After-tax cash flow, $ |
|---|---|---|---|---|---|
| 0 | -10,000 | | | | -10,000 |
| 1 to 5 | 3000 | 2000 | 1000 | 460 | 2540 |
| 6 to 10 | 3000 | | 3000 | 1380 | 1620 |
| Total | 20,000 | 10,000 | 20,000 | 9200 | 10,800 |

Recall that the before-tax return for this investment is 27.3% and the after-tax return using a 10-year, SL writeoff is 16.1%. Let us assume that the IRS guideline life for this investment classification is 5 years, even though the time period of economic interest is 10 years. Table 6.3 shows the accounting steps required to compute the after-tax cash flow on the basis of a 5-year writeoff and a 10-year economic life.

Notice from the table that the $10,000 writeoff occurs during the first 5 years at a uniform rate of $2000/year. During these years, the taxable savings are $1000 ($3000 less $2000), the tax is 46% of $1000 ($460), and the after-tax cash flow is $3000 less the tax. During the next 5 years, no writeoff occurs and the entire $3000 savings is taxable. With a constant tax rate, the total tax paid over the 10 years is unchanged from the preceding example, but the shortened writeoff has caused a portion of the earlier years' taxes to be deferred. The resulting after-tax cash flow diagram is shown in Figure 6.5.

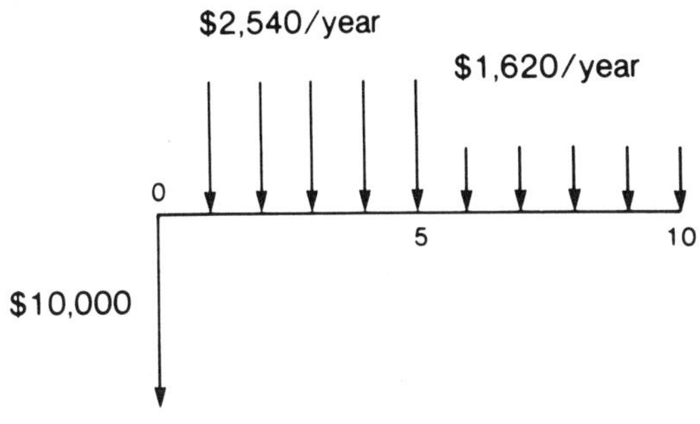


**Figure 6.5 Example 3: 5-year writeoff and 10-year economic life**

To find the return on investment, we may solve the cash flow equation for i:

$$\$10000 = \$1620 \left(\frac{P}{A}\right)_{N=10} + \$920 \left(\frac{P}{A}\right)_{N=5}$$

The result is *18.6%*, which may be compared to 16.1% for 10-year writeoff. The ROI has been increased 2.5% simply because of a reduction in writeoff life.

Let us continue this example by examining an extended writeoff of *20* years. Table 6.4 shows the corresponding intermediate steps.

In computing our tax during years 11 to 20 (after termination of the project), we assumed that the taxpayer would apply the continuing tax deductions to tax liabilities from other projects. Notice that the totals are all identical to the 5-year writeoff, but the taxes must now be paid earlier. It may be shown that the resulting return is *14.0%* (compared to 16.1% for 10-year writeoff).

To conclude Example 3, let us now assume that the total expenditure may be expensed during the first year. Research and development is an example of an expenditure that often produces income for many years but may be expensed during the year of occurrence. Table 6.5 presents the resulting cash flows. The

**Table 6.4 Cash flow and accounting table for Example 3 with 20-year writeoff and 10-year economic life**

| End-of-year | Before-tax cash flow, $ | Writeoff $ | Taxable income, $ | Tax @ 46% | After-tax cash flow, $ |
|---|---|---|---|---|---|
| 0 | -10,000 | | | | -10,000 |
| 1 to 10 | 3000 | 500 | 2500 | 1150 | 1850 |
| 11 to 20 | | 500 | -500 | -230 | 230 |
| Total | 20,000 | 10,000 | 20,000 | 9200 | 10,800 |

**Table 6.5 Cash flow and accounting table for Example 3 with 1-year writeoff and 10-year economic life**

| End-of-year | Before-tax cash flow, $ | Expense $ | Taxable income, $ | Tax @ 46% | After-tax cash flow, $ |
|---|---|---|---|---|---|
| 0 | -10,000 | | | | -10,000 |
| 1 | 3000 | 10,000 | -7000 | -3220 | 6220 |
| 2 to 10 | 3000 | | 3000 | 1300 | 1620 |
| Total | 20,000 | 10,000 | 20,000 | 9200 | 10,800 |

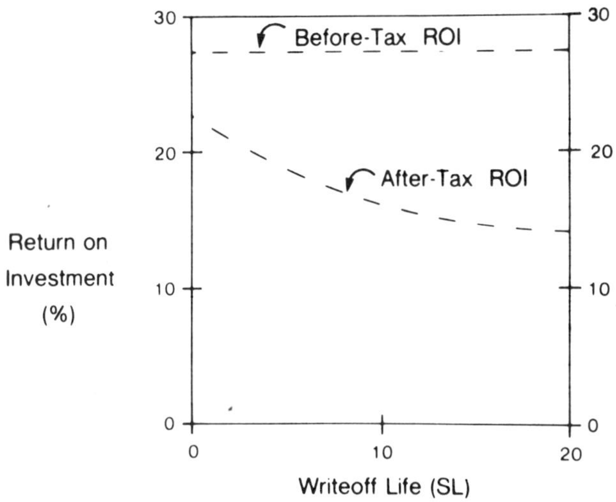


**Figure 6.6 Impact of writeoff period (SL) on after-tax ROI**

return on the expensed investment may be found to be 22.7%. The impact of writeoff life on investment return for Example 3 is illustrated graphically by the summary results shown in Figure 6.6.

From the examples on writeoff method and writeoff life, it should now be clear that governmental tax policy influences corporate and private investment strategies. Tax policy may stifle the taxpayers' motivation to produce income or it may provide powerful incentives for productive investment. Accelerated writeoff methods, shortened writeoff lives, additional first year depreciation allowances, and investment tax credits are examples of governmental incentives to renew our nation's productivity capability. Tax credits will be discussed later in this chapter.

## Depletion Allowance

The tax deduction for exhaustible natural resources and timber is termed *depletion.* Natural resources include oil and gas wells, geothermal deposits, mines, and other mineral deposits.

There are two methods for determining depletion allowance: (1) cost depletion, and (2) percentage depletion. If the taxpayer is allowed a choice, the method yielding the larger tax deduction for that particular year should be used.

*Cost depletion* is a writeoff based on units consumed. The cost depletion allowance for a given year generally is computed by first determining a cost per unit and multiplying by the number of units sold during that year. The cost per unit is obtained by dividing the value of the mineral property by the total number of recoverable units (tons, barrels, board-feet, etc.). Determination of the total number of recoverable units is primarily an engineering problem, to be found in accordance with prevailing industry standards. The value of the property is the original cost less all depletion previously allowed or allowable.

The *percentage depletion* for a given year is a certain percentage of the gross income from the property during that year. It may not exceed 50% of the taxable income computed without depletion. Percentage depletion for oil and gas is no longer allowed except for certain domestic production by small producers, nor is the percentage method permitted for timber.

Gross income from oil and gas property is the market value at the wellhead. Gross income from mining property is the gross income attributable to extraction and mining treatment of the mineral or ore. Some of the more common depletion percentages are given in Table 6.6.

*Example 4: Timber* Assume that you bought a timber tract for $100,000 including land worth $60,000 and an estimated 1 million board-feet of standing timber. During the tax year, you sold 100,000 board-feet of cut timber.

Timber depletion is computed only by the cost method, since percentage depletion is not permitted. Since the cost of the timber alone is $40,000 for an estimated 1 million board-feet, the depletion unit is $.04/board-feet. The depletion allowance for the tax year would be 100,000 × $.04 or $4000.

*Example 5: Gravel Pit* Assume that you have purchased 40 acres of land with subsurface gravel along the Mississippi River. Gravel capacity has been estimated from borings as 400,000 tons. Purchase price was $200,000.

**Table 6.6 Percentage depletion allowances for selected exhaustible natural resources**

| Deposit | Percentage |
|---|---|
| Geothermal | 22[a] |
| Sulphur, uranium and U.S. deposits of asbestos, lead, zinc, nickel, and mica | 22 |
| U.S. deposits of gold, silver, copper, iron ore, and oil shale | 15 |
| Coal and sodium chloride | 10 |
| Clay for bricks and sewer pipe | 7½ |
| Gravel, sand, stone, drainage tile clay | 5 |
| Most other minerals | 14 |

[a] Decreasing in steps to 15% after 1983.

During the first year of operation, 10,000 tons of gravel are mined and sold at a production cost of \$20,000. Market price at the pit was \$5/ton.

The first year's depletion allowance computed using the cost method is \$5000:

$$\text{Unit depletion} = \frac{\text{First cost} - \text{accumulated depletion}}{\text{remaining capacity}}$$

$$= \frac{\$200{,}000 - 0}{400{,}000} = \$.50/\text{ton}$$

$$\text{Total depletion} = (\$.50/\text{ton})(10{,}000 \text{ tons}) = \$5000$$

Using the percentage method, the first year's allowance is \$2500:

$$\text{Gross income} = (10{,}000 \text{ tons})(\$5/\text{ton}) = \$50{,}000$$

Depletion = Fixed percentage of gross income (not to exceed half of the net taxable income before depletion)

$$= (5\%)(\$50{,}000) = \$2500$$

$$\text{Maximum} = \frac{(\$50{,}000 - \$20{,}000)}{2} = \$15{,}000 > \$2500$$

For the first year, the depletion allowance using the cost method (\$5000) is larger than that for the percentage method (\$2500).

## Tax Credits

Unlike a tax deduction, a tax credit is subtracted directly from the taxes otherwise due. For example, an individual who donates \$50 to the campaign of a political candidate may claim half of the donation as a tax credit, thereby reducing his or her taxes by \$25. As an alternative, the individual may claim a \$50 itemized deduction. For a taxpayer in the 40% tax bracket who itemizes deductions, the contribution would reduce taxes by \$20.

Because tax credits are subtracted from the bottom line, they exert a powerful influence on individual and corporate decision making. Tax credits available to the individual include:

Residential energy credits

Credit for political contributions

Credit for the elderly

Investment credit

Federal gasoline and oil credit (nonhighway use)

Probably the most important corporate tax credit is the investment credit. The purpose of the investment tax credit is to encourage modernization and expansion of American industry through greater capital investment in machines and equipment. For machinery and equipment having a useful life of at least 3 years and used as an integral part of manufacturing, production, or extraction, a 10% investment credit is allowed for the first year of the investment.* The full investment is subject to the investment tax credit if the writeoff life is at least 7 years. Lesser writeoff lives reduce the tax credit. Claiming an investment tax credit does not reduce the depreciation permitted.

The National Energy Act of 1978 provides an additional 10% investment credit for business investment in energy property, for example, recycling, solar, and wind equipment. The Energy Act also provides two residential energy investment credits to encourage energy conservation and the development of alternate energy sources for the home:

A 10% credit on the first $2000 invested in qualifying home energy-conserving expenditures, for example, storm windows and insulation

A 30% credit on the first $2000 and 20% on the next $8000 spent on renewable energy source equipment for the home, for example, solar collectors and geothermal wells

The following example illustrates the dramatic impact of the investment tax credit on an investment's economic return.

***Example 6: Investment Tax Credit*** The proposed project involves the installation of refractory cloth curtains on a metal, heat-treating furnace in a foundry. The expenditure qualifies for both the general and the energy tax credits.

| *Project* | | *Furnace curtains* |
|---|---|---|
| First cost | = | $70,000 |
| Life | = | 7 years |
| Salvage | = | 0 |
| Before-tax savings | = | $20,000 |
| Tax rate | = | 46% |

The cash flow computation table is shown in Table 6.7. Notice that 20% of the first cost is underwritten by the government through the two 10% tax

*The amount of the investment tax credit is subject to periodic Congressional review.

Table 6.7 Accounting and cash flow table illustrating investment tax credits (Example 6)

| End-of-year | Before-tax cash flow, $ | Writeoff (SL), $ | Taxable income, $ | Tax (w/o credit), $ | 10% + 10% credits, $ | After-tax cash flow, $ |
|---|---|---|---|---|---|---|
| 0 | -70,000 | | | | | -70,000 |
| 1 | | | | | 14,000 | 14,000 |
| 1 to 7 | 20,000 | 10,000 | 10,000 | 4600 | | 15,400 |

credits. Tax credits totaling $14,000 are subtracted directly from the company's tax bill. The return on investment may be found by solving the after-tax cash flow pattern for i:

$$\$70{,}000 = \$15{,}400\left(\frac{P}{A}\right)_7 + \$14{,}000\left(\frac{P}{F}\right)_1$$

Solving the equation for i, the after-tax return with the investment tax credit may be found to be 18.4%. It can be shown that without the credits, the return on the identical investment is 12.1%.

The use of accelerated writeoff would further increase the after-tax return on this investment. Also, this investment qualifies for an additional 20%, first-year writeoff for tangible personal property, as described in detail in IRS Publication 534, *Tax Information on Depreciation.*

# 7 The Effect of Borrowing on Investment Economics

## Equity Versus Debt Financing

We have seen that the principal objective of economic analysis is to assist the decision maker in allocating scarce capital. Four common methods have been described for comparing alternatives on the basis of their projected economic impacts: present worth, annual worth, benefit/cost, and return on investment (ROI). With all four approaches, the alternatives are first expressed in terms of their expected cash flows over the desired life span. The cash flows are identified from a particular viewpoint or perspective, that is, as money entering or leaving a particular well-defined economic system.

Up to this point we have not concerned ourselves with the source of the investor's capital. In a sense, we have implied that all of the capital originated from within the investor's economic system, in other words, from the investor's own equity. This approach, often described as the *100% equity* viewpoint, is invaluable for assessing the economics of the total investment. But in those instances where a substantial fraction of the capital is borrowed from others, it is also important to examine the economics from the standpoint of the *investor's share* of the capital. Obviously these two viewpoints and resulting cash flows may be quite different. From the investor's eyes, the borrowed funds are cash flows entering the system while the debt repayment are cash flows leaving the system.

An examination of the impact of borrowing on investment economics is especially important when the investment represents a major capital expenditure for the investor, an expenditure that in itself requires additional borrowing. For the small investor, such an expenditure may be only a few thousand dollars, while for a large corporation a major investment may involve a hundred million dollars.

## The Concept of Leverage

Usually investment funds are borrowed at an interest rate that is fixed for the life of the loan, for example, industrial bonds and bank notes. When the percentage return on the total investment exceeds the interest rate charged on the debt, the return on the investor's equity is magnified or *levered* in favor of the investor.

To introduce this important concept of leverage, consider a corporation with $100 million in bond indebtedness costing 8% interest. If the firm's net earnings before payment of bond interest is $10 million, then $2 million remains for the owners after payment of the $8 million in bond interest.

Let us now assume that the before-bond earnings increase $1 million or 10%, from $10 million to $11 million, with no change in bond indebtedness. The owners' share of the earnings now increases from $2 million to $3 million, *a change of 50%.*

Obviously the owners' return is strongly levered by the fixed bond indebtedness. But the lever works both ways. While it benefits the owner during times of increasing earnings, it also amplifies any drop in earnings. Notice that a relatively modest 10% decrease in before-bond earnings is magnified to a precipitous 50% drop in the owners' earnings. A 20% decrease in before-bond earnings wipes out the owner's earnings entirely.

Levered investments combine greater potential for higher profits with greater vulnerability to unexpected business downturns. Higher leverage is most attractive when the probability of an unexpected downturn is minimal.

To illustrate leverage in the absence of tax effects, we will reconsider the before-tax analysis of Example 2 in Chapter 6 by assuming varying amounts of borrowed capital:

### *Example 1: 50% and 80% Borrowed Capital*

*Proposed investment*

| | |
|---|---|
| First cost | = $10,000 |
| Salvage | = 0 |
| Life | = 10 yr |
| Before-tax savings | = $3000/yr |

You may recall from the earlier example that the before-tax return on this investment is 27.3% based on the total investment of $10000, that is, from the 100% equity viewpoint.

Let us now assume that $5000 or half of the investment will be borrowed at 10% interest, leaving a balance of $5000 or 50% owner's equity. Also assume that debt interest is to be paid at the end of each year with the principal due at the end of 10 years. Since the yearly interest is a cash flow from the standpoint of the owner, the owner's net yearly before-tax cash flow becomes \$2500 (\$3000 less \$500 interest). In addition, the loan must be repaid at the end of the tenth year. The owner's cash flow diagram is shown in Figure 7.1. To find the owner's before-tax return on investment, we solve the corresponding equation for i:

$$\$5000 = \$2500 \left(\frac{P}{A}\right) - \$5000 \left(\frac{P}{F}\right)$$

The resulting ROI is about *48%* compared to 27.3% without the loan.

We now continue the example by increasing the leverage to 80% borrowed funds:

*80% Borrowed funds*

First cost = \$10000 (\$8000 borrowed and \$2000 equity)

Interest = \$800/yr

Principal = \$8000 at the end of the tenth yr

The owner's cash flow diagram corresponding to the 80% loan is shown in Figure 7.2. The before-tax return on the owner's \$2000 may be computed from the corresponding equation:

$$\$2000 = \$2200 \left(\frac{P}{A}\right) - \$8000 \left(\frac{P}{F}\right)$$

The owner's ROI is about *110%*!

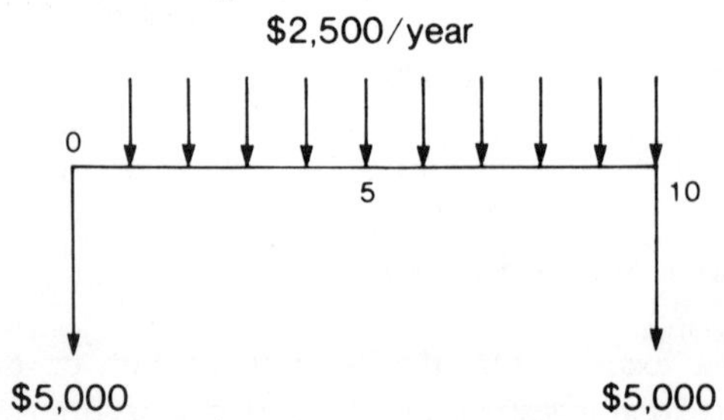


**Figure 7.1 Example: Owner's before-tax cash flow on \$10,000 total investment with 50% borrowed capital**

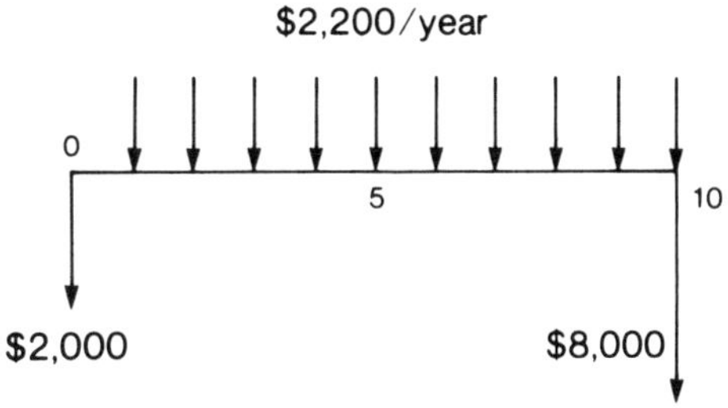


Figure 7.2 Example: Owner's before-tax cash flow with 80% borrowed capital

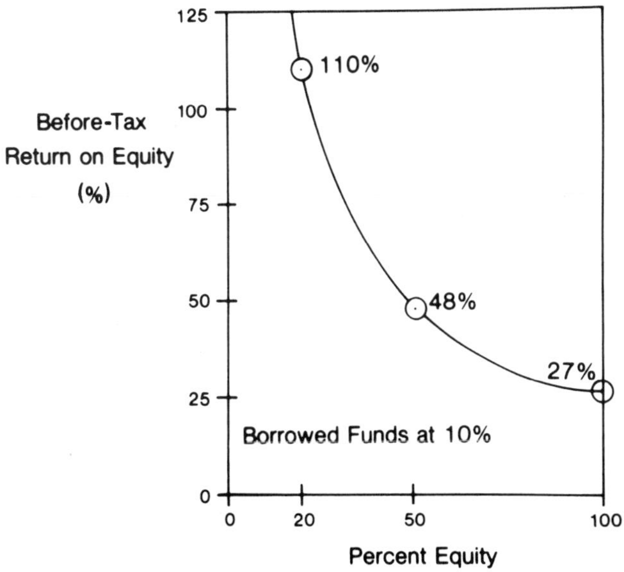


Figure 7.3 Effect of borrowing at 10% on owner's before-tax return on equity (Example)

The dramatic impact of borrowing on the before-tax investment return for this example is summarized graphically in Figure 7.3.

## Income Tax and Leverage

The phenomenon of leverage is a result of two factors: (1) simple algebraic partitioning of an investment into two segments, debt and equity, at different interest rates, and (2) dissimilar tax treatments of debt and equity. The algebraic weighting effect was illustrated in the before-tax analysis of Example 1. The tax effects will be illustrated in this section by extending Example 1 to an after-tax analysis.

Existing accounting standards and Internal Revenue Codes discriminate against equity capital in favor of debt capital. The interest on debt capital is tax-deductible, so the real cost to a corporate taxpayer in the 50% bracket is only half of its pretax cost. No comparable deduction is permitted for the owner's equity capital. For example, common and preferred stocks are considered to be equity, and consequently the payment of stock dividends is not an allowable tax-deductible expense. In addition to the favored treatment given to debt interest, an investment in depreciable property tax may be depreciated in its entirety by the legal owner, even if the investment is heavily debt financed.

The tax-deductibility of interest and depreciation offers a positive incentive for borrowing when investing in depreciable property. The incentive is particularly attractive to taxpayers in higher tax brackets. To illustrate income tax effects operating in conjunction with debt leverage, we will now extend Example 1 to an after-tax analysis with debt financing.

*Example 2: Tax and Debt Leverage Effects* For tax purposes, we will assume an income tax rate of 46% and straightline (SL) depreciation:

| | |
|---|---|
| First cost | = $10,000 |
| Salvage | = 0 |
| Life, writeoff and economic | = 10 years |
| Before-tax savings | = $3000/yr |
| Tax rate | = 46% |
| Depreciation method | = SL |

You may recall that the after-tax, 100% equity case was considered in Example 2 of Chapter 6. The after-tax return on the total investment was found to be 16.1%.

Let us now assume that half of the initial investment is borrowed at 10%, with interest due at the end of each year and the principal at the end of 10 years. The resulting accounting and cash flow data are shown in Table 7.1.

Notice from Table 7.1 that the owner's initial investment is reduced to $5000, with the $5000 balance supplied by the loan. The owner pays an additional $5000 upon repayment of the loan principal at the end of the tenth year. Straight-line depreciation is taken on the entire $10,000 investment, $1000 each year for 10 years. Remember that depreciation is not a cash flow. In contrast, the $500 interest is definitely a cash flow that must be paid by the owner. Both interest and the depreciation writeoff are tax-deductible, leaving

**Table 7.1 Accounting and After-tax Table for Example 2 Using 50% Borrowed Funds**

| Year | Before-tax cash flow, $ | Writeoff $ | Interest, $ | Taxable savings, $ | Tax @ 46% $ | After-tax cash flow, $ |
|---|---|---|---|---|---|---|
| 0 | -5000 | | | | | -5000 |
| 1 to 10 | 3000 | $1000 | 500 | 1500 | 690 | 1810 |
| 10 | -5000 | | | | | -5000 |

a taxable savings of $1500 ($3000 less writeoff and interest). The resulting income tax (also a cash flow for the owner) is 46% of $1500, or $690. Subtracting the tax and interest cash flows, we find that the owner's yearly after-tax cash flow is $1810 ($3000 less $500 interest less $690 taxes), together with the $5000 loan repayment at the end of the tenth year.

The right-hand column of Table 7.1 represents the after-tax cash flow from the viewpoint of the owner's $5000 investment. The corresponding ROI is *32%*, compared to 16.1% without the loan.

The example may be extended to an 80% loan, with the owner paying $2000 initially and then the $8000 loan repayment in 10 years. The resulting annual after-tax cash flow is $1648 ($3000 savings less $800 interest less $552 tax). Again, depreciation writeoff of the entire $10000 investment is permitted, in spite of the fact that the owner's share is only $2000. And the interest is also tax-deductible as an expense. The after-tax ROI on the owner's $2000 is *81%*.

Throughout this example, the investment opportunity was identical. The differences lie in the source of the investment capital and the resulting differences in the tax treatment. The dramatic influence of tax and debt leverage effects is summarized graphically in Figure 7.4, together with the earlier before-tax results.

## Financial Equity/Debt Ratios

Because of the debt and tax leverage advantages, most U.S. corporations obtain investment capital through debt financing as well as owner equity.

Various capitalization ratios provide financial measures of a firm's indebtedness and its resulting leverage. As an example, the *bond ratio* is the ratio of long-term debt in the form of bonds to the total capital funds (bonds plus owner's tangible equity: preferred stock, common stock, additional paid-in capital, and retained earnings). Another popular financial ratio that expresses the relationship between capital contributed by creditors and that contributed

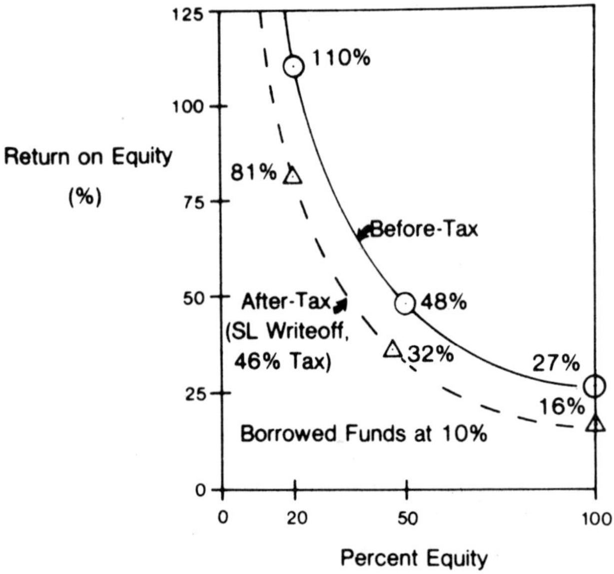


**Figure 7.4 Effect of borrowing at 10% on owner's after-tax return on equity (Example)**

**Table 7.2 Representative Financial Data for Selected U.S. Industry/Utility Categories**[a]

| Industry category | Bond ratio | Average income tax rate, % | Percentage earned on total capital | Percentage earned on equity |
|---|---|---|---|---|
| Air transport | .51 | 18 | 11 | 17 |
| Building industry | .31 | 46 | 12 | 15 |
| Chemical (basic) | .31 | 42 | 11 | 14 |
| Computer/data processing | .11 | 46 | 16 | 18 |
| Electric utilities (east) | .51 | 34 | 7 | 10 |
| Metal fabricating | .33 | 46 | 11 | 14 |
| Mobil home | .10 | 48 | 13 | 14 |
| Petroleum (integrated) | .29 | 61 | 11 | 13 |
| Steel (general) | .25 | 47 | 10 | 12 |
| Telecommunication | .47 | 37% | 8% | 12% |

[a] Data excerpted from *The Value Line Investment Survey, Part 3: Ratings and Reports,* Arnold Bernhard & Co., Inc., New York (1979).

by owners is *debt/worth,* the ratio of total liabilities (accounts payable, taxes payable, short-term notes, long-term bonds) to tangible net worth.

Capitalization ratios vary considerably from industry to industry. Regulated fields such as utilities usually have a heavier debt load than industrial companies since their revenue streams have in the past been more predictable. Recent financial data for selected U.S. industry/utility categories are shown in Table 7.2, including bond ratio, average income tax rate, and approximate percentage earned on total capital and owner's equity or net worth.

# 8

# Three Case Studies

To summarize and highlight the fundamentals of economic decision making, we conclude with three applications to practical problems followed by a final summary.

The first case illustrates the application of economic decision theory to a classic engineering investment problem. While the economic results are valuable to the decision maker, the choice ultimately rests on noneconomic, technical, judgmental factors. The analysis is before taxes.

The second case involves capital budgeting at the highest corporate level, with and without the financial constraint of limited capital. With unlimited capital, the investment alternatives are totally independent. Under the financial constraint of limited capital, the alternatives become financially dependent and must be arranged into mutually exclusive groupings for ultimate comparison.

In the third case, we examine the cash flow impact and after-tax economics of an investment situation frequently encountered by engineering consultants and by individual investors. Accelerated depreciation and borrowed capital are included in the example.

## Case 1: Air Pollution Controls

An Ohio Valley utility is being pressed by the Environmental Protection Agency (EPA) to reduce the particulate emissions from one of its coal-fired stations. The station is equipped with six boiler units each supplying a 130,000 kilowatt turbine set. Although each unit is presently equipped with an electrostatic precipitator (ESP), they are not operating at design efficiency due to changes in coal quality. Consequently the emissions do not satisfy federal codes for particulates.

Two proposals have been prepared for correcting the deficiency:

Install a new baghouse between the existing precipitator and chimney, or

Replace the ESP with a new, more efficient ESP unit

Both the baghouse and the new precipitator are designed to meet federal standards. Cost estimates have been prepared and are presented in Table 8.1. Using the utility's minimum attractive return i* of 12%, the data in Table 8.1, and a 15-year life for both alternatives, we may compare the equivalent annual costs of the two alternatives:

Baghouse

$$A = \$77000 + \$760000\left(\frac{A}{P}\right)_{15} = \$188{,}600$$

ESP

$$A = \$48800 + \$735000\left(\frac{A}{P}\right)_{15} = \$156{,}700$$

Notice that the electrostatic precipitator is less expensive both from the standpoint of its first cost and annual operating cost. Its total equivalent annual cost

**Table 8.1 Cost and accounting estimate for the baghouse and precipitator in Example 1**

| | Baghouse, $ | ESP, $ |
|---|---|---|
| Purchase cost of device | 372,000 | 340,000 |
| Purchase cost, fan and motor | 8,000 | 4,000 |
| Structures and ductwork | 2,000 | 3,000 |
| Controls | 30,000 | 40,000 |
| Installation | 340,000 | 340,000 |
| Startup | 8,000 | 8,000 |
| Total first cost | 760,000 | 735,000 |
| Annual power cost | 28,800/yr | 6,200/yr |
| Maintenance | 36,600/yr | 31,000/yr |
| Ash disposal | 11,600/yr | 11,600/yr |
| Total annual cost | 77,000/yr | 48,800/yr |

is about $31,900 lower than that of the baghouse. The ESP is the clear economic choice. Nevertheless, in this situation the engineers felt the economic advantage of ESP was insufficient to overcome certain noneconomic advantages of the baghouse, particularly its ability to perform up to design specifications under varying operating conditions. The baghouse was recommended.

### Case 2: Corporate Capital Budgeting

According to corporate policy for TransAmerican Industries, all requests for capital expenditures exceeding $50,000 are submitted by the respective operating divisions to the capital budgeting committee comprised of the board chairman, president, comptroller, and two vice-presidents. The alternatives are then grouped into major categories according to their level of risk and uncertainty, for example, a new warehouse versus a new product line.

The investment alternatives for the low-risk group are listed by operating division in Table 8.2. Annual savings are assumed uniform and salvage values have been neglected for simplicity.

If the minimum attractive return rate i* for the low-risk group is 15% and *capital is unlimited*, find the economically acceptable alternatives.

The alternatives are functionally independent. Since no limit exists on capital, they are also financially independent. Consequently, they may be treated as independent alternatives wherein any or all may be selected.

**Table 8.2 Investment alternatives for Example 2 listed by operating division**

| Investment alternatives | First cost, $10,000s | Life, yr | Annual savings, $10,000s |
|---|---|---|---|
| Minneapolis | | | |
| 1. Insulate plant roof | 100 | 40 | 20 |
| 2. Upgrade truck fleet | 100 | 5 | 25 |
| Des Moines | | | |
| 3. Relocate warehouse near I-80 | 400 | 40 | 70 |
| Baltimore | | | |
| 4. Purchase heat recovery equipment | 200 | 10 | 40 |
| 5. Install automatic process controls | 300 | 20 | 50 |
| 6. Install sprinkler system | 100 | 40 | 18 |
| San Jose | | | |
| 7. Purchase water recycling equipment | 200 | 20 | 40 |

The corporate objective is to maximize the net equivalent annual savings. Selecting the annual worth method of solution, we first determine the equivalent annual savings for each alternative. For alternative 1, the equivalent annual savings may be found to be \$4.90 (expressed in \$10,000s):

$$A = \$20 - \$100\left(\frac{A}{P}\right)_{40} = \$20 - \$100\,(.1506) = \$4.90$$

The equivalent annual savings for the other alternatives may be computed in similar fashion. The results are summarized in Table 8.3. By examining the table, we find that for the case of unlimited capital, the net corporate savings will be maximized by selecting all alternatives except number 2. All except number 2 show positive equivalent savings and therefore yield a return of at least 15%. The total equivalent savings for the six acceptable alternatives is \$278,000.

To illustrate the case of financial dependence, let us now assume that the capital expended for this low-risk category may not exceed \$*4 million.* To proceed, we first arrange the alternatives into all possible combinations within this capital ceiling. Since only one combination may be selected, we may treat the various combinations as mutually exclusive alternatives, choosing the one combination that yields the highest equivalent annual savings. The combinations with their corresponding savings are listed in Table 8.4. The economic choice that maximizes savings is the combination of alternatives 1, 6, and 7, for a total equivalent annual savings of \$158,000 at an i* of 15%.

Since the investment opportunities that appear attractive at an i* of 15% far exceed the available capital in this case, the budgeting committee would be wise to ration the capital by raising the i* writeoff. Increasing i* will squeeze out additional projects until only the best remain. An essential characteristic of a successful capital budgeting program, yet one that is frequently slighted, is the aggressive solicitation and formulation of new investment opportunities. The

**Table 8.3 Equivalent annual savings for the seven alternatives in Example 2**

| Alternative | First cost, \$10,000s | Equivalent annual savings, \$10,000s |
|---|---|---|
| 1 | 100 | 4.9 |
| 2 | 100 | -4.8 |
| 3 | 400 | 9.8 |
| 4 | 200 | 0.1 |
| 5 | 300 | 2.1 |
| 6 | 100 | 2.9 |
| 7 | 200 | 8.0 |

Table 8.4 Mutually exclusive combinations of alternatives satisfying the $4 million maximum capital constraint (Case 2) and their total equivalent annual savings

| Combination of alternatives | Total investment, $10,000s | Total annual savings, $10,000s |
|---|---|---|
| 1 | 100 | 4.9 |
| 3 | 400 | 9.8 |
| 4 | 200 | 0.1 |
| 5 | 300 | 2.1 |
| 6 | 100 | 2.9 |
| 7 | 200 | 8.0 |
| 1,4 | 300 | 5.0 |
| 1,6 | 200 | 7.8 |
| 1,7 | 300 | 12.9 |
| 4,6 | 300 | 3.0 |
| 4,7 | 400 | 8.1 |
| 5,1 | 400 | 7.0 |
| 5,6 | 400 | 5.0 |
| 6,7 | 300 | 10.9 |
| 1,4,6 | 400 | 7.9 |
| 1,6,7 | 400 | 15.8 |

firm able to generate opportunities far in excess of available capital is in the enviable position of requiring a higher return on new investments.

## Case 3: Professional Office Complex

An engineering and architectural firm is considering the purchase of a small professional office complex for $120,000 with $20,000 cash down payment and a $100,000 mortgage. The firm anticipates that the practice will outgrow the complex in about five years at which time it will be sold. Sum-of-the-years-digits (SOYD) capital writeoff would be utilized under the Class Life Asset Depreciation Range (CLADR) system (45-year guideline life), and the mortgage would be repaid in uniform payments ($875/month totaling $10,500/year) over a period of about 30 years. Investment information and estimates have been prepared and are summarized in Table 8.5.

Table 8.5 Information on the professional office complex in Example 3

| | |
|---|---|
| Purchase price | \$120,000 (\$100,000 bldg and \$20,000 land) |
| Project life | 5 yr |
| Est. Sale price in 5 yr, net | \$145,000 (based on 4% annual appreciation) |
| Depreciation, CLADR system | |
| Life | 45 yr (guideline) |
| Method | SOYD |
| Salvage | zero[a] |
| Gross rentals | \$15,000/yr[b] |
| Operating expenses, maintenance, taxes, etc. | \$3000/yr |
| Tax bracket | 40% |
| Mortgage | \$100,000 @ 10% (\$875/month, or \$10,500/yr for about 30 years) |

[a]Zero salvage permitted for CLADR system.
[b]Gross rental includes rent that engineering firm would be obliged to pay elsewhere if office complex is not purchased.

As a close approximation, assume that the mortgage is paid yearly rather than monthly. At the end of the first year, the mortgage interest due is 10% of \$100,000 or \$10,000. Since the mortgage payment is \$10,500, the difference of \$500 is applied to the principal, reducing it to \$99,500. The second year's interest is 10% of \$99,500.

The after-tax cash flow table may now be completed and is shown in Table 8.6. The after-tax return on investment may be found by solving the equation for the cash flows shown in the right-hand column of the table:

$$\$20{,}000 = \$2439\left(\frac{P}{F}\right)_1 + \$2380\left(\frac{P}{F}\right)_2 + 2320\left(\frac{P}{F}\right)_3 + \$2257\left(\frac{P}{F}\right)_4 + (\$2192 + \$29{,}743)\left(\frac{P}{F}\right)_5$$

The result is about 18%.

It is quite possible that the firm may want to examine the sensitiveness of the result to an error in one or several of the input estimates, for example, sale price. One approach is to simply repeat the computation for return-on-investment (ROI) using a different value of the sale price. Another approach is to use the minimum attractive rate i* to compute the sale price needed to "breakeven,"

**Table 8.6 Accounting and after-tax cash flow table for professional office complex in Example 3**

| Year | Gross rentals less expenses, $ | Writeoff, $ | Mortgage payment/interest, $ | Taxable income, $ | Tax @ 40%, $ | After-tax cash flow, $ |
|---|---|---|---|---|---|---|
| 0 | -20000 | | | | | -20000 |
| 1 | 12000 | 4348[a] | 10500/10000 | -2348[b] | -939 | 2439[c] |
| 2 | 12000 | 4251 | 10500/9950 | -2201 | -880 | 2380 |
| 3 | 12000 | 4155 | 10500/9895 | -2050 | -820 | 2320 |
| 4 | 12000 | 4058 | 10500/9835 | -1893 | -757 | 2257 |
| 5 | 12000 | 3961 | 10500/9768 | -1729 | -692 | 2192 |
| 5 | 48052[d] | | | 45773[e] | 18309[f] | 29743 |

[a]SOYD writeoff for the first year is 45/1035 times the building value of $100,000.
[b]Taxable income for years 1 to 5 is $12,000 less writeoff and interest.
[c]After-tax cash flow for years 1 to 5 is $12,000 less mortgage payment and tax.
[d]Net before-tax cash flow from sale of property is the net sale value ($145,000) less the unpaid loan principal ($96,948).
[e]Taxable gain upon sale of property is the net sale value ($145,000) less the undepreciated balance or book value ($99,227).
[f]For simplicity in the example, the gain upon sale ($45,773) was taxed as ordinary income. A portion of the gain may qualify as long-term capital gain, which is subject to a substantially lower tax. Complete details are given in IRS Publication 334, *Tax Guide for Small Business*.

that is, to yield i* and no more. A breakeven sensitivity analysis would probably be desirable on the income input (gross rentals less expenses) as well as sale price, since these are the two major uncertainties in the investment.

Notice that the office complex would show an accounting loss (negative taxable income) each of the 5 years, thereby sheltering a portion of the income from the engineering practice, even though the after-tax cash flows for all 5 years following the purchase are positive.

## Final Summary

The necessity of making decisions—choosing among alternatives—is a fact of life brought about by our scarce resources. In the broadest possible sense, decision making is simply the allocation of scarce resources to fulfill one's needs and objectives.

When the primary resource is capital, when an investment in the near future must be balanced against future profits or returns, when the central question is "Will this investment pay?", then engineering economics is the single most valuable quantitative tool available to the decision maker today.

Whether one describes the approach as engineering economy, life cycle costing, economic decision analysis, or managerial economics, it must be founded on a few fundamental principles:

Define the economic system by establishing the appropriate viewpoint.

Identify all reasonable alternatives.

Forecast the outcome of each alternative in terms of its cash flows across the system boundary.

Select a minimum attractive rate of return i* reflective of the capital supply and demand and the risk level of the investments.

Compare the alternatives on an equivalent basis—present or annual worth, benefit/cost, or ROI—using a common project lifetime and reconciling differences in cash flow timing by applying the interest concept.

The strength of engineering economy lies in its ability to quantify alternatives when the major decision factors are expressible in terms of dollars. The singular weakness of engineering economy is its inadequacy in dealing with nonmonetary factors, factors that cannot be fairly represented or perhaps not represented at all by money.

Even in those instances where money is an adequate common denominator, the results of the economic analysis are no more dependable than the cash flow projections on which the analysis is built. Nor does the analysis assure consideration of the proper and best alternatives. In the capital budgeting process, the strategic, long-range planning function of identifying attractive alternatives is the crucial activity in the decision-making endeavor. Economic analysis is an essential support rather than a substitute for the ultimate creative, judgemental act of human decision making.

# Selected Readings

## General Texts

American Telephone and Telegraph Company *Engineering Economy: A Manager's Guide to Economic Decision Making.* 3rd ed. McGraw-Hill, New York (1977).

Barish, N. N. and S. Kaplan *Economic Analysis for Engineering and Managerial Decision Making,* 2nd ed. McGraw-Hill, New York (1978).

Bierman, H. Jr. and S. Smidt *The Capital Budgeting Decision: Economic Analysis and Financing of Investment Projects.* 3rd ed. The MacMillan Company, New York (1971).

Canada, J. R. and J. A. White Jr. *Capital Investment Decision Analysis.* Prentice-Hall, Englewood Cliffs, N.J. (1980).

Dean, J., *Managerial Economics,* Prentice-Hall, Englewood Cliffs, N.J. (1971).

Fabrycky, W. J. and G. J. Thuesen *Economic Decision Analysis,* 2nd ed. (1980). Prentice-Hall, Englewood Cliffs, N.J.

Grant, E. L., W. G. Ireson and R. S. Leavenworth *Principles of Engineering Economy.* 6th ed. Ronald Press, New York (1976).

Newman, D. G. *Engineering Economic Analysis,* Engineering Press, San Jose, Calif. (1976).

Riggs, J. L. *Engineering Economics,* McGraw-Hill, New York (1977).

Smith, G. W. *Engineering Economy: Analysis of Capital Expenditures.* 2nd ed. Iowa State University Press, Ames (1973).

Tarquin, A. J. and L. T. Blank *Engineering Economy: A Behavioral Approach,* McGraw-Hill, New York (1976).

Taylor, G. A. *Managerial and Engineering Economy.* 2nd ed. D. Van Nostrand, New York (1975).

Thuesen, H. G. , W. J. Fabrycky and G. J. Thuesen *Engineering Economy.* 5th ed. Prentice-Hall, Englewood Cliffs, N.J. (1977).

White, J. A., M. H. Agee and K. E. Case *Principles of Engineering Economic Analysis,* John Wiley & Sons, Somerset, N.J. (1977).

### Methodology

Brown, K. C. The rate of return on selected investment projects. *Journal of Business,* September 1978.

Conley, B. C. The value of human life in the demand for safety. *American Economic Review,* March 1976.

Gibbons, J. E. and S. Rushmore, Using total project analysis to compete for investment capital. *The Appraisal Journal,* October 1975.

Heenan, D. A. and Robert Addleman, Quantitative techniques for today's decision makers. *Harvard Business Review,* May/June 1976.

Jeynes, P. H. Fundamentals of accounting/economic financial evaluation. *The Engineering Economist,* Summer 1975.

Kulonda, D. J. Replacement analysis with unequal lives: The study period method. *The Engineering Economist,* Spring 1978.

White, J. A., K. E. Case and M. H. Agee Rate of return: An explicit reinvestment approach, *1976 Spring Annual Conference Proceedings,* American Institute of Industrial Engineers.

Zussman, R. Predict the rate of return on net investments: The communications payoff. *Data Communications,* July 1979.

### Capital Budgeting

Bradley, S. P., R. S. Frank and S. C. Frey, Jr. Determining the appropriate discount rates in pure capital rationing. *Decision Sciences,* July 1978.

Chopp, C. C. Proper discount rate for economy studies, *1978 Iowa State University Regulatory Conference Proceedings.*

Hsu, J. I. S. A survey of equipment replacement practices. *Managerial Planning,* March/April 1975.

Klammer, T. The association of capital budgeting techniques with firm performance. *The Accounting Review,* April 1973.

Maier, S. F. and J. H. Vander Weide Capital budgeting in the decentralized firm. *Management Science,* December 1976.

Miller, M. H. and C. W. Upton Leasing, buying and the cost of capital services. *The Journal of Finance,* June 1976.

Naylor, T. H. and D. R. Gattis Corporate planning models. *California Management Review,* Summer 1976.

Porwal, L. S. and S. S. Singhui A comparative study of capital expenditure evaluation techniques. *Long Range Planning,* October 1978.

Schoeffler, S., R. D. Buzzell and D. F. Heany Impact of strategic planning on profit performance. *Harvard Business Review,* March/April 1974.

Searby, F. W. Return to return on investment. *Harvard Business Review,* March/April 1975.

Thuessen, G. J. Selecting a discount rate: Let's help the user, *1975 Spring Annual Conference Proceedings,* American Institute of Industrial Engineers.

## Income Tax

Eisner, R. and P. J. Lawler Tax policy and investment. *The American Economic Review,* March 1975.

Grant, E. L. Some lessons from sixty years of treatment of depreciation and related matters for income tax purposes in the United States, *1973 Spring Annual Conference Proceedings,* American Institute of Industrial Engineers.

Smith, G. W. and R. D. Chartier Optimizing the switch from DDB to SOYD. *Industrial Engineer,* April 1973.

U.S. Department of The Treasury, Internal Revenue Service, *Your Federal Income Tax* Publication 17 (annual).

U.S. Department of the Treasury, Internal Revenue Service, *Tax Guide for Small Business* Publication 334 (annual).

U.S. Department of the Treasury, Internal Revenue Service, *Tax Information on Depreciation* Publication 534 (annual).

U.S. Department of the Treasury, Internal Revenue Service, *Tax Information on Investment Credit* Publication 572 (annual).

U.S. Department of the Treasury, Internal Revenue Service, *Catalog and Quick Index to Tax Publications* Publication 900 (annual).

## Debt/Equity

Arditti, F. D. and J. M. Pinkerton The valuation and cost of capital of the levered firm with growth opportunities. *The Journal of Finance,* March 1978.

Donaldson, G. New Framework for Corporate Debt Policy. *Harvard Business Review,* September/October 1978.

Gordon, M. J. and C. I. Gould The cost of equity capital: A consideration. *Journal of Finance,* June 1978.

Gordon, M. J. and P. H. Halpern Cost of capital for a division of a firm. *The Journal of Finance,* September 1974.

Lasman, D. A. and R. L. Weil Adjusting the Debt-Equity Ratio. *Financial Analyst's Journal,* September/October 1978.

Robert Morris Associates *RMA Annual Statement Studies,* Philadelphia, Penn. (annual).

Arnold Bernhard and Co. *The Value Line Investment Survey, Part 3: Ratings and Reports,* New York (annual).

Troy, L. *Almanac of Business and Industrial Financial Ratios,* Prentice-Hall, Englewood Cliffs, N.J. (annual).

## Dealing with Risk and Uncertainty

Fleischer, G. A., ed. *Risk and Uncertainty: Non-Deterministic Decision Making in Engineering Economy,* American Institute of Industrial Engineers, Norcross, Ga. (1975).

Greenberg, E., W. J. Marshall and J. B. Yawity The technology of risk and return. *American Economic Review,* June 1978.

Hammonds, T. M. Discounting for risk. *Quarterly Review of Economics and Business,* Autumn 1971.

Hertz, D. Risk analysis in capital investments. *Harvard Business Review,* September/October 1979.

Martin, W. B., Jr. A risk analysis rate-of-return model for evaluating income producing real estate investments. *The Appraisal Journal,* July 1978.

Reisman, A. and A. K. Rao *Discounted Cash Flow Analysis: Stochastic Extensions,* American Institute of Industrial Engineers, Norcross, Ga. (1973).

Rose, L. M. *Engineering Investment Decisions: Planning Under Uncertainty,* Elsevier Scientific Publishing Company, Amsterdam (1976).

Taylor, B. W., III and K. R. Davis Checking risks in capital investments. *Industrial Engineering,* December 1978.

Vedder, J. N. Treatment of differential costs and benefits under risk. *Decision Sciences,* April 1978.

# Appendix 1
## Additional Interest Equations

P/F equations for N and i

$$N = \frac{\log_e (F/P)}{\log_e (1 + i)} \tag{1}$$

$$i = \left(\frac{F}{P}\right)^{1/N} - 1 \tag{2}$$

P/A equation for N

$$N = -\frac{\log_e (1 - i\,P/A)}{\log_e (1 + i)} \tag{3}$$

F/A equation for N

$$N = \frac{\log_e (i\,F/A + 1)}{\log_e (1 + i)} \tag{4}$$

# Appendix 2
# Compound Interest Tables

## Formulas for Calculating Compound Interest Factors

Compound amount factor, single payment $\frac{F}{P} = (1 + i)^N$

Present worth factor, single payment $\frac{P}{F} = \frac{1}{(1 + i)^N}$

Present worth factor, uniform series $\frac{P}{A} = \frac{(1 + i)^N - 1}{i(1 + i)^N}$

Capital recovery factor $\frac{A}{P} = \frac{i(1 + i)^N}{(1 + i)^N - 1}$

Sinking fund factor $\frac{A}{F} = \frac{i}{(1 + i)^N - 1}$

Compound amount factor, uniform series $\frac{F}{A} = \frac{(1 + i)^N - 1}{i}$

0.25% COMPOUND INTEREST FACTORS

| N | F/P | P/F | F/A | A/F | P/A | A/P | N |
|---|---|---|---|---|---|---|---|
| 1 | 1.0025 | 0.9975 | 1.0000 | 1.0000 | 0.9975 | 1.0025 | 1 |
| 2 | 1.0050 | 0.9950 | 2.0025 | 0.4994 | 1.9925 | 0.5019 | 2 |
| 3 | 1.0075 | 0.9925 | 3.0075 | 0.3325 | 2.9851 | 0.3350 | 3 |
| 4 | 1.0100 | 0.9901 | 4.0150 | 0.2491 | 3.9751 | 0.2516 | 4 |
| 5 | 1.0126 | 0.9876 | 5.0251 | 0.1990 | 4.9627 | 0.2015 | 5 |
| 6 | 1.0151 | 0.9851 | 6.0376 | 0.1656 | 5.9478 | 0.1681 | 6 |
| 7 | 1.0176 | 0.9827 | 7.0527 | 0.1418 | 6.9305 | 0.1443 | 7 |
| 8 | 1.0202 | 0.9802 | 8.0704 | 0.1239 | 7.9107 | 0.1264 | 8 |
| 9 | 1.0227 | 0.9778 | 9.0905 | 0.1100 | 8.8885 | 0.1125 | 9 |
| 10 | 1.0253 | 0.9753 | 10.1133 | 0.0989 | 9.8639 | 0.1014 | 10 |
| 11 | 1.0278 | 0.9729 | 11.1385 | 0.0898 | 10.8368 | 0.0923 | 11 |
| 12 | 1.0304 | 0.9705 | 12.1664 | 0.0822 | 11.8073 | 0.0847 | 12 |
| 13 | 1.0330 | 0.9681 | 13.1968 | 0.0758 | 12.7753 | 0.0783 | 13 |
| 14 | 1.0356 | 0.9656 | 14.2298 | 0.0703 | 13.7410 | 0.0728 | 14 |
| 15 | 1.0382 | 0.9632 | 15.2654 | 0.0655 | 14.7042 | 0.0680 | 15 |
| 16 | 1.0408 | 0.9608 | 16.3035 | 0.0613 | 15.6650 | 0.0638 | 16 |
| 17 | 1.0434 | 0.9584 | 17.3443 | 0.0577 | 16.6235 | 0.0602 | 17 |
| 18 | 1.0460 | 0.9561 | 18.3876 | 0.0544 | 17.5795 | 0.0569 | 18 |
| 19 | 1.0486 | 0.9537 | 19.4336 | 0.0515 | 18.5332 | 0.0540 | 19 |
| 20 | 1.0512 | 0.9513 | 20.4822 | 0.0488 | 19.4845 | 0.0513 | 20 |
| 21 | 1.0538 | 0.9489 | 21.5334 | 0.0464 | 20.4334 | 0.0489 | 21 |
| 22 | 1.0565 | 0.9466 | 22.5872 | 0.0443 | 21.3800 | 0.0468 | 22 |
| 23 | 1.0591 | 0.9442 | 23.6437 | 0.0423 | 22.3241 | 0.0448 | 23 |
| 24 | 1.0618 | 0.9418 | 24.7028 | 0.0405 | 23.2660 | 0.0430 | 24 |
| 25 | 1.0644 | 0.9395 | 25.7646 | 0.0388 | 24.2055 | 0.0413 | 25 |
| 26 | 1.0671 | 0.9371 | 26.8290 | 0.0373 | 25.1426 | 0.0398 | 26 |
| 27 | 1.0697 | 0.9348 | 27.8961 | 0.0358 | 26.0774 | 0.0383 | 27 |
| 28 | 1.0724 | 0.9325 | 28.9658 | 0.0345 | 27.0099 | 0.0370 | 28 |
| 29 | 1.0751 | 0.9301 | 30.0382 | 0.0333 | 27.9400 | 0.0358 | 29 |
| 30 | 1.0778 | 0.9278 | 31.1133 | 0.0321 | 28.8679 | 0.0346 | 30 |
| 31 | 1.0805 | 0.9255 | 32.1911 | 0.0311 | 29.7934 | 0.0336 | 31 |
| 32 | 1.0832 | 0.9232 | 33.2716 | 0.0301 | 30.7166 | 0.0326 | 32 |
| 33 | 1.0859 | 0.9209 | 34.3547 | 0.0291 | 31.6375 | 0.0316 | 33 |
| 34 | 1.0886 | 0.9186 | 35.4406 | 0.0282 | 32.5561 | 0.0307 | 34 |
| 35 | 1.0913 | 0.9163 | 36.5292 | 0.0274 | 33.4724 | 0.0299 | 35 |
| 36 | 1.0941 | 0.9140 | 37.6206 | 0.0266 | 34.3865 | 0.0291 | 36 |
| 37 | 1.0968 | 0.9118 | 38.7146 | 0.0258 | 35.2982 | 0.0283 | 37 |
| 38 | 1.0995 | 0.9095 | 39.8114 | 0.0251 | 36.2077 | 0.0276 | 38 |
| 39 | 1.1023 | 0.9072 | 40.9109 | 0.0244 | 37.1149 | 0.0269 | 39 |
| 40 | 1.1050 | 0.9050 | 42.0132 | 0.0238 | 38.0199 | 0.0263 | 40 |
| 50 | 1.1330 | 0.8826 | 53.1887 | 0.0188 | 46.9462 | 0.0213 | 50 |
| 60 | 1.1616 | 0.8609 | 64.6467 | 0.0155 | 55.6524 | 0.0180 | 60 |
| 70 | 1.1910 | 0.8396 | 76.3944 | 0.0131 | 64.1439 | 0.0156 | 70 |
| 80 | 1.2211 | 0.8189 | 88.4392 | 0.0113 | 72.4260 | 0.0138 | 80 |
| 90 | 1.2520 | 0.7987 | 100.7885 | 0.0099 | 80.5038 | 0.0124 | 90 |
| 100 | 1.2836 | 0.7790 | 113.4500 | 0.0088 | 88.3825 | 0.0113 | 100 |
| 110 | 1.3161 | 0.7598 | 126.4316 | 0.0079 | 96.0669 | 0.0104 | 110 |
| 120 | 1.3494 | 0.7411 | 139.7414 | 0.0072 | 103.5618 | 0.0097 | 120 |
| 180 | 1.5674 | 0.6380 | 226.9727 | 0.0044 | 144.8055 | 0.0069 | 180 |
| 240 | 1.8208 | 0.5492 | 328.3020 | 0.0030 | 180.3109 | 0.0055 | 240 |
| 300 | 2.1150 | 0.4728 | 446.0078 | 0.0022 | 210.8765 | 0.0047 | 300 |
| 360 | 2.4568 | 0.4070 | 582.7369 | 0.0017 | 237.1894 | 0.0042 | 360 |

0.50% COMPOUND INTEREST FACTORS

| N | F/P | P/F | F/A | A/F | P/A | A/P | N |
|---|---|---|---|---|---|---|---|
| 1 | 1.0050 | 0.9950 | 1.0000 | 1.0000 | 0.9950 | 1.0050 | 1 |
| 2 | 1.0100 | 0.9901 | 2.0050 | 0.4988 | 1.9851 | 0.5038 | 2 |
| 3 | 1.0151 | 0.9851 | 3.0150 | 0.3317 | 2.9702 | 0.3367 | 3 |
| 4 | 1.0202 | 0.9802 | 4.0301 | 0.2481 | 3.9505 | 0.2531 | 4 |
| 5 | 1.0253 | 0.9754 | 5.0503 | 0.1980 | 4.9259 | 0.2030 | 5 |
| 6 | 1.0304 | 0.9705 | 6.0755 | 0.1646 | 5.8964 | 0.1696 | 6 |
| 7 | 1.0355 | 0.9657 | 7.1059 | 0.1407 | 6.8621 | 0.1457 | 7 |
| 8 | 1.0407 | 0.9609 | 8.1414 | 0.1228 | 7.8230 | 0.1278 | 8 |
| 9 | 1.0459 | 0.9561 | 9.1821 | 0.1089 | 8.7791 | 0.1139 | 9 |
| 10 | 1.0511 | 0.9513 | 10.2280 | 0.0978 | 9.7304 | 0.1028 | 10 |
| 11 | 1.0564 | 0.9466 | 11.2792 | 0.0887 | 10.6770 | 0.0937 | 11 |
| 12 | 1.0617 | 0.9419 | 12.3356 | 0.0811 | 11.6189 | 0.0861 | 12 |
| 13 | 1.0670 | 0.9372 | 13.3972 | 0.0746 | 12.5562 | 0.0796 | 13 |
| 14 | 1.0723 | 0.9326 | 14.4642 | 0.0691 | 13.4887 | 0.0741 | 14 |
| 15 | 1.0777 | 0.9279 | 15.5365 | 0.0644 | 14.4166 | 0.0694 | 15 |
| 16 | 1.0831 | 0.9233 | 16.6142 | 0.0602 | 15.3399 | 0.0652 | 16 |
| 17 | 1.0885 | 0.9187 | 17.6973 | 0.0565 | 16.2586 | 0.0615 | 17 |
| 18 | 1.0939 | 0.9141 | 18.7858 | 0.0532 | 17.1728 | 0.0582 | 18 |
| 19 | 1.0994 | 0.9096 | 19.8797 | 0.0503 | 18.0824 | 0.0553 | 19 |
| 20 | 1.1049 | 0.9051 | 20.9791 | 0.0477 | 18.9874 | 0.0527 | 20 |
| 21 | 1.1104 | 0.9006 | 22.0840 | 0.0453 | 19.8880 | 0.0503 | 21 |
| 22 | 1.1160 | 0.8961 | 23.1944 | 0.0431 | 20.7841 | 0.0481 | 22 |
| 23 | 1.1216 | 0.8916 | 24.3104 | 0.0411 | 21.6757 | 0.0461 | 23 |
| 24 | 1.1272 | 0.8872 | 25.4320 | 0.0393 | 22.5629 | 0.0443 | 24 |
| 25 | 1.1328 | 0.8828 | 26.5591 | 0.0377 | 23.4456 | 0.0427 | 25 |
| 26 | 1.1385 | 0.8784 | 27.6919 | 0.0361 | 24.3240 | 0.0411 | 26 |
| 27 | 1.1442 | 0.8740 | 28.8304 | 0.0347 | 25.1980 | 0.0397 | 27 |
| 28 | 1.1499 | 0.8697 | 29.9745 | 0.0334 | 26.0677 | 0.0384 | 28 |
| 29 | 1.1556 | 0.8653 | 31.1244 | 0.0321 | 26.9330 | 0.0371 | 29 |
| 30 | 1.1614 | 0.8610 | 32.2800 | 0.0310 | 27.7941 | 0.0360 | 30 |
| 31 | 1.1672 | 0.8567 | 33.4414 | 0.0299 | 28.6508 | 0.0349 | 31 |
| 32 | 1.1730 | 0.8525 | 34.6086 | 0.0289 | 29.5033 | 0.0339 | 32 |
| 33 | 1.1789 | 0.8482 | 35.7817 | 0.0279 | 30.3515 | 0.0329 | 33 |
| 34 | 1.1848 | 0.8440 | 36.9606 | 0.0271 | 31.1955 | 0.0321 | 34 |
| 35 | 1.1907 | 0.8398 | 38.1454 | 0.0262 | 32.0354 | 0.0312 | 35 |
| 36 | 1.1967 | 0.8356 | 39.3361 | 0.0254 | 32.8710 | 0.0304 | 36 |
| 37 | 1.2027 | 0.8315 | 40.5328 | 0.0247 | 33.7025 | 0.0297 | 37 |
| 38 | 1.2087 | 0.8274 | 41.7354 | 0.0240 | 34.5299 | 0.0290 | 38 |
| 39 | 1.2147 | 0.8232 | 42.9441 | 0.0233 | 35.3531 | 0.0283 | 39 |
| 40 | 1.2208 | 0.8191 | 44.1588 | 0.0226 | 36.1722 | 0.0276 | 40 |
| 50 | 1.2832 | 0.7793 | 56.6452 | 0.0177 | 44.1428 | 0.0227 | 50 |
| 60 | 1.3489 | 0.7414 | 69.7700 | 0.0143 | 51.7256 | 0.0193 | 60 |
| 70 | 1.4178 | 0.7053 | 83.5661 | 0.0120 | 58.9394 | 0.0170 | 70 |
| 80 | 1.4903 | 0.6710 | 98.0677 | 0.0102 | 65.8023 | 0.0152 | 80 |
| 90 | 1.5666 | 0.6383 | 113.3109 | 0.0088 | 72.3313 | 0.0138 | 90 |
| 100 | 1.6467 | 0.6073 | 129.3337 | 0.0077 | 78.5426 | 0.0127 | 100 |
| 110 | 1.7309 | 0.5777 | 146.1759 | 0.0068 | 84.4518 | 0.0118 | 110 |
| 120 | 1.8194 | 0.5496 | 163.8793 | 0.0061 | 90.0735 | 0.0111 | 120 |
| 180 | 2.4541 | 0.4075 | 290.8187 | 0.0034 | 118.5035 | 0.0084 | 180 |
| 240 | 3.3102 | 0.3021 | 462.0409 | 0.0022 | 139.5808 | 0.0072 | 240 |
| 300 | 4.4650 | 0.2240 | 692.9940 | 0.0014 | 155.2069 | 0.0064 | 300 |
| 360 | 6.0226 | 0.1660 | 1004.5150 | 0.0010 | 166.7916 | 0.0060 | 360 |

## 0.75% COMPOUND INTEREST FACTORS

| N | F/P | P/F | F/A | A/F | P/A | A/P | N |
|---|---|---|---|---|---|---|---|
| 1 | 1.0075 | 0.9926 | 1.0000 | 1.0000 | 0.9926 | 1.0075 | 1 |
| 2 | 1.0151 | 0.9852 | 2.0075 | 0.4981 | 1.9777 | 0.5056 | 2 |
| 3 | 1.0227 | 0.9778 | 3.0226 | 0.3308 | 2.9556 | 0.3383 | 3 |
| 4 | 1.0303 | 0.9706 | 4.0452 | 0.2472 | 3.9261 | 0.2547 | 4 |
| 5 | 1.0381 | 0.9633 | 5.0756 | 0.1970 | 4.8894 | 0.2045 | 5 |
| 6 | 1.0459 | 0.9562 | 6.1136 | 0.1636 | 5.8456 | 0.1711 | 6 |
| 7 | 1.0537 | 0.9490 | 7.1595 | 0.1397 | 6.7946 | 0.1472 | 7 |
| 8 | 1.0616 | 0.9420 | 8.2132 | 0.1218 | 7.7366 | 0.1293 | 8 |
| 9 | 1.0696 | 0.9350 | 9.2748 | 0.1078 | 8.6716 | 0.1153 | 9 |
| 10 | 1.0776 | 0.9280 | 10.3443 | 0.0967 | 9.5996 | 0.1042 | 10 |
| 11 | 1.0857 | 0.9211 | 11.4219 | 0.0876 | 10.5207 | 0.0951 | 11 |
| 12 | 1.0938 | 0.9142 | 12.5076 | 0.0800 | 11.4349 | 0.0875 | 12 |
| 13 | 1.1020 | 0.9074 | 13.6014 | 0.0735 | 12.3423 | 0.0810 | 13 |
| 14 | 1.1103 | 0.9007 | 14.7034 | 0.0680 | 13.2430 | 0.0755 | 14 |
| 15 | 1.1186 | 0.8940 | 15.8137 | 0.0632 | 14.1370 | 0.0707 | 15 |
| 16 | 1.1270 | 0.8873 | 16.9323 | 0.0591 | 15.0243 | 0.0666 | 16 |
| 17 | 1.1354 | 0.8807 | 18.0593 | 0.0554 | 15.9050 | 0.0629 | 17 |
| 18 | 1.1440 | 0.8742 | 19.1947 | 0.0521 | 16.7792 | 0.0596 | 18 |
| 19 | 1.1525 | 0.8676 | 20.3387 | 0.0492 | 17.6468 | 0.0567 | 19 |
| 20 | 1.1612 | 0.8612 | 21.4912 | 0.0465 | 18.5080 | 0.0540 | 20 |
| 21 | 1.1699 | 0.8548 | 22.6524 | 0.0441 | 19.3628 | 0.0516 | 21 |
| 22 | 1.1787 | 0.8484 | 23.8223 | 0.0420 | 20.2112 | 0.0495 | 22 |
| 23 | 1.1875 | 0.8421 | 25.0010 | 0.0400 | 21.0533 | 0.0475 | 23 |
| 24 | 1.1964 | 0.8358 | 26.1885 | 0.0382 | 21.8891 | 0.0457 | 24 |
| 25 | 1.2054 | 0.8296 | 27.3849 | 0.0365 | 22.7188 | 0.0440 | 25 |
| 26 | 1.2144 | 0.8234 | 28.5903 | 0.0350 | 23.5422 | 0.0425 | 26 |
| 27 | 1.2235 | 0.8173 | 29.8047 | 0.0336 | 24.3595 | 0.0411 | 27 |
| 28 | 1.2327 | 0.8112 | 31.0282 | 0.0322 | 25.1707 | 0.0397 | 28 |
| 29 | 1.2420 | 0.8052 | 32.2609 | 0.0310 | 25.9759 | 0.0385 | 29 |
| 30 | 1.2513 | 0.7992 | 33.5029 | 0.0298 | 26.7751 | 0.0373 | 30 |
| 31 | 1.2607 | 0.7932 | 34.7542 | 0.0288 | 27.5683 | 0.0363 | 31 |
| 32 | 1.2701 | 0.7873 | 36.0148 | 0.0278 | 28.3557 | 0.0353 | 32 |
| 33 | 1.2796 | 0.7815 | 37.2849 | 0.0268 | 29.1371 | 0.0343 | 33 |
| 34 | 1.2892 | 0.7757 | 38.5646 | 0.0259 | 29.9128 | 0.0334 | 34 |
| 35 | 1.2989 | 0.7699 | 39.8538 | 0.0251 | 30.6827 | 0.0326 | 35 |
| 36 | 1.3086 | 0.7641 | 41.1527 | 0.0243 | 31.4468 | 0.0318 | 36 |
| 37 | 1.3185 | 0.7585 | 42.4614 | 0.0236 | 32.2053 | 0.0311 | 37 |
| 38 | 1.3283 | 0.7528 | 43.7798 | 0.0228 | 32.9581 | 0.0303 | 38 |
| 39 | 1.3383 | 0.7472 | 45.1082 | 0.0222 | 33.7053 | 0.0297 | 39 |
| 40 | 1.3483 | 0.7416 | 46.4465 | 0.0215 | 34.4469 | 0.0290 | 40 |
| 50 | 1.4530 | 0.6883 | 60.3943 | 0.0166 | 41.5664 | 0.0241 | 50 |
| 60 | 1.5657 | 0.6387 | 75.4241 | 0.0133 | 48.1734 | 0.0208 | 60 |
| 70 | 1.6872 | 0.5927 | 91.6201 | 0.0109 | 54.3046 | 0.0184 | 70 |
| 80 | 1.8180 | 0.5500 | 109.0725 | 0.0092 | 59.9944 | 0.0167 | 80 |
| 90 | 1.9591 | 0.5104 | 127.8790 | 0.0078 | 65.2746 | 0.0153 | 90 |
| 100 | 2.1111 | 0.4737 | 148.1445 | 0.0068 | 70.1746 | 0.0143 | 100 |
| 110 | 2.2749 | 0.4396 | 169.9823 | 0.0059 | 74.7219 | 0.0134 | 110 |
| 120 | 2.4514 | 0.4079 | 193.5143 | 0.0052 | 78.9417 | 0.0127 | 120 |
| 180 | 3.8380 | 0.2605 | 378.4058 | 0.0026 | 98.5934 | 0.0101 | 180 |
| 240 | 6.0092 | 0.1664 | 667.8869 | 0.0015 | 111.1450 | 0.0090 | 240 |
| 300 | 9.4084 | 0.1063 | 1121.1219 | 0.0009 | 119.1616 | 0.0084 | 300 |
| 360 | 14.7306 | 0.0679 | 1830.7435 | 0.0005 | 124.2819 | 0.0080 | 360 |

1.00% COMPOUND INTEREST FACTORS

| N | F/P | P/F | F/A | A/F | P/A | A/P | N |
|---|---|---|---|---|---|---|---|
| 1 | 1.0100 | 0.9901 | 1.0000 | 1.0000 | 0.9901 | 1.0100 | 1 |
| 2 | 1.0201 | 0.9803 | 2.0100 | 0.4975 | 1.9704 | 0.5075 | 2 |
| 3 | 1.0303 | 0.9706 | 3.0301 | 0.3300 | 2.9410 | 0.3400 | 3 |
| 4 | 1.0406 | 0.9610 | 4.0604 | 0.2463 | 3.9020 | 0.2563 | 4 |
| 5 | 1.0510 | 0.9515 | 5.1010 | 0.1960 | 4.8534 | 0.2060 | 5 |
| 6 | 1.0615 | 0.9420 | 6.1520 | 0.1625 | 5.7955 | 0.1725 | 6 |
| 7 | 1.0721 | 0.9327 | 7.2135 | 0.1386 | 6.7282 | 0.1486 | 7 |
| 8 | 1.0829 | 0.9235 | 8.2857 | 0.1207 | 7.6517 | 0.1307 | 8 |
| 9 | 1.0937 | 0.9143 | 9.3685 | 0.1067 | 8.5660 | 0.1167 | 9 |
| 10 | 1.1046 | 0.9053 | 10.4622 | 0.0956 | 9.4713 | 0.1056 | 10 |
| 11 | 1.1157 | 0.8963 | 11.5668 | 0.0865 | 10.3676 | 0.0965 | 11 |
| 12 | 1.1268 | 0.8874 | 12.6825 | 0.0788 | 11.2551 | 0.0888 | 12 |
| 13 | 1.1381 | 0.8787 | 13.8093 | 0.0724 | 12.1337 | 0.0824 | 13 |
| 14 | 1.1495 | 0.8700 | 14.9474 | 0.0669 | 13.0037 | 0.0769 | 14 |
| 15 | 1.1610 | 0.8613 | 16.0969 | 0.0621 | 13.8651 | 0.0721 | 15 |
| 16 | 1.1726 | 0.8528 | 17.2579 | 0.0579 | 14.7179 | 0.0679 | 16 |
| 17 | 1.1843 | 0.8444 | 18.4304 | 0.0543 | 15.5623 | 0.0643 | 17 |
| 18 | 1.1961 | 0.8360 | 19.6147 | 0.0510 | 16.3983 | 0.0610 | 18 |
| 19 | 1.2081 | 0.8277 | 20.8109 | 0.0481 | 17.2260 | 0.0581 | 19 |
| 20 | 1.2202 | 0.8195 | 22.0190 | 0.0454 | 18.0456 | 0.0554 | 20 |
| 21 | 1.2324 | 0.8114 | 23.2392 | 0.0430 | 18.8570 | 0.0530 | 21 |
| 22 | 1.2447 | 0.8034 | 24.4716 | 0.0409 | 19.6604 | 0.0509 | 22 |
| 23 | 1.2572 | 0.7954 | 25.7163 | 0.0389 | 20.4558 | 0.0489 | 23 |
| 24 | 1.2697 | 0.7876 | 26.9735 | 0.0371 | 21.2434 | 0.0471 | 24 |
| 25 | 1.2824 | 0.7798 | 28.2432 | 0.0354 | 22.0232 | 0.0454 | 25 |
| 26 | 1.2953 | 0.7720 | 29.5256 | 0.0339 | 22.7952 | 0.0439 | 26 |
| 27 | 1.3082 | 0.7644 | 30.8209 | 0.0324 | 23.5596 | 0.0424 | 27 |
| 28 | 1.3213 | 0.7568 | 32.1291 | 0.0311 | 24.3164 | 0.0411 | 28 |
| 29 | 1.3345 | 0.7493 | 33.4504 | 0.0299 | 25.0658 | 0.0399 | 29 |
| 30 | 1.3478 | 0.7419 | 34.7849 | 0.0287 | 25.8077 | 0.0387 | 30 |
| 31 | 1.3613 | 0.7346 | 36.1327 | 0.0277 | 26.5423 | 0.0377 | 31 |
| 32 | 1.3749 | 0.7273 | 37.4941 | 0.0267 | 27.2696 | 0.0367 | 32 |
| 33 | 1.3887 | 0.7201 | 38.8690 | 0.0257 | 27.9897 | 0.0357 | 33 |
| 34 | 1.4026 | 0.7130 | 40.2577 | 0.0248 | 28.7027 | 0.0348 | 34 |
| 35 | 1.4166 | 0.7059 | 41.6603 | 0.0240 | 29.4086 | 0.0340 | 35 |
| 36 | 1.4308 | 0.6989 | 43.0769 | 0.0232 | 30.1075 | 0.0332 | 36 |
| 37 | 1.4451 | 0.6920 | 44.5076 | 0.0225 | 30.7995 | 0.0325 | 37 |
| 38 | 1.4595 | 0.6852 | 45.9527 | 0.0218 | 31.4847 | 0.0318 | 38 |
| 39 | 1.4741 | 0.6784 | 47.4123 | 0.0211 | 32.1630 | 0.0311 | 39 |
| 40 | 1.4889 | 0.6717 | 48.8864 | 0.0205 | 32.8347 | 0.0305 | 40 |
| 50 | 1.6446 | 0.6080 | 64.4632 | 0.0155 | 39.1961 | 0.0255 | 50 |
| 60 | 1.8167 | 0.5504 | 81.6697 | 0.0122 | 44.9550 | 0.0222 | 60 |
| 70 | 2.0068 | 0.4983 | 100.6763 | 0.0099 | 50.1685 | 0.0199 | 70 |
| 80 | 2.2167 | 0.4511 | 121.6715 | 0.0082 | 54.8882 | 0.0182 | 80 |
| 90 | 2.4486 | 0.4084 | 144.8633 | 0.0069 | 59.1609 | 0.0169 | 90 |
| 100 | 2.7048 | 0.3697 | 170.4814 | 0.0059 | 63.0289 | 0.0159 | 100 |
| 110 | 2.9878 | 0.3347 | 198.7797 | 0.0050 | 66.5305 | 0.0150 | 110 |
| 120 | 3.3004 | 0.3030 | 230.0387 | 0.0043 | 69.7005 | 0.0143 | 120 |
| 180 | 5.9958 | 0.1668 | 499.5802 | 0.0020 | 83.3217 | 0.0120 | 180 |
| 240 | 10.8926 | 0.0918 | 989.2554 | 0.0010 | 90.8194 | 0.0110 | 240 |
| 300 | 19.7885 | 0.0505 | 1878.8466 | 0.0005 | 94.9466 | 0.0105 | 300 |
| 360 | 35.9496 | 0.0278 | 3494.9641 | 0.0003 | 97.2183 | 0.0103 | 360 |

1.25% COMPOUND INTEREST FACTORS

| N | F/P | P/F | F/A | A/F | P/A | A/P | N |
|---|---|---|---|---|---|---|---|
| 1 | 1.0125 | 0.9877 | 1.0000 | 1.0000 | 0.9877 | 1.0125 | 1 |
| 2 | 1.0252 | 0.9755 | 2.0125 | 0.4969 | 1.9631 | 0.5094 | 2 |
| 3 | 1.0380 | 0.9634 | 3.0377 | 0.3292 | 2.9265 | 0.3417 | 3 |
| 4 | 1.0509 | 0.9515 | 4.0756 | 0.2454 | 3.8781 | 0.2579 | 4 |
| 5 | 1.0641 | 0.9398 | 5.1266 | 0.1951 | 4.8178 | 0.2076 | 5 |
| 6 | 1.0774 | 0.9282 | 6.1907 | 0.1615 | 5.7460 | 0.1740 | 6 |
| 7 | 1.0909 | 0.9167 | 7.2680 | 0.1376 | 6.6627 | 0.1501 | 7 |
| 8 | 1.1045 | 0.9054 | 8.3589 | 0.1196 | 7.5681 | 0.1321 | 8 |
| 9 | 1.1183 | 0.8942 | 9.4634 | 0.1057 | 8.4623 | 0.1182 | 9 |
| 10 | 1.1323 | 0.8832 | 10.5817 | 0.0945 | 9.3455 | 0.1070 | 10 |
| 11 | 1.1464 | 0.8723 | 11.7139 | 0.0854 | 10.2178 | 0.0979 | 11 |
| 12 | 1.1608 | 0.8615 | 12.8604 | 0.0778 | 11.0793 | 0.0903 | 12 |
| 13 | 1.1753 | 0.8509 | 14.0211 | 0.0713 | 11.9302 | 0.0838 | 13 |
| 14 | 1.1900 | 0.8404 | 15.1964 | 0.0658 | 12.7706 | 0.0783 | 14 |
| 15 | 1.2048 | 0.8300 | 16.3863 | 0.0610 | 13.6005 | 0.0735 | 15 |
| 16 | 1.2199 | 0.8197 | 17.5912 | 0.0568 | 14.4203 | 0.0693 | 16 |
| 17 | 1.2351 | 0.8096 | 18.8111 | 0.0532 | 15.2299 | 0.0657 | 17 |
| 18 | 1.2506 | 0.7996 | 20.0462 | 0.0499 | 16.0295 | 0.0624 | 18 |
| 19 | 1.2662 | 0.7898 | 21.2968 | 0.0470 | 16.8193 | 0.0595 | 19 |
| 20 | 1.2820 | 0.7800 | 22.5630 | 0.0443 | 17.5993 | 0.0568 | 20 |
| 21 | 1.2981 | 0.7704 | 23.8450 | 0.0419 | 18.3697 | 0.0544 | 21 |
| 22 | 1.3143 | 0.7609 | 25.1431 | 0.0398 | 19.1306 | 0.0523 | 22 |
| 23 | 1.3307 | 0.7515 | 26.4574 | 0.0378 | 19.8820 | 0.0503 | 23 |
| 24 | 1.3474 | 0.7422 | 27.7881 | 0.0360 | 20.6242 | 0.0485 | 24 |
| 25 | 1.3642 | 0.7330 | 29.1354 | 0.0343 | 21.3573 | 0.0468 | 25 |
| 26 | 1.3812 | 0.7240 | 30.4996 | 0.0328 | 22.0813 | 0.0453 | 26 |
| 27 | 1.3985 | 0.7150 | 31.8809 | 0.0314 | 22.7963 | 0.0439 | 27 |
| 28 | 1.4160 | 0.7062 | 33.2794 | 0.0300 | 23.5025 | 0.0425 | 28 |
| 29 | 1.4337 | 0.6975 | 34.6954 | 0.0288 | 24.2000 | 0.0413 | 29 |
| 30 | 1.4516 | 0.6889 | 36.1291 | 0.0277 | 24.8889 | 0.0402 | 30 |
| 31 | 1.4698 | 0.6804 | 37.5807 | 0.0266 | 25.5693 | 0.0391 | 31 |
| 32 | 1.4881 | 0.6720 | 39.0504 | 0.0256 | 26.2413 | 0.0381 | 32 |
| 33 | 1.5067 | 0.6637 | 40.5386 | 0.0247 | 26.9050 | 0.0372 | 33 |
| 34 | 1.5256 | 0.6555 | 42.0453 | 0.0238 | 27.5605 | 0.0363 | 34 |
| 35 | 1.5446 | 0.6474 | 43.5709 | 0.0230 | 28.2079 | 0.0355 | 35 |
| 36 | 1.5639 | 0.6394 | 45.1155 | 0.0222 | 28.8473 | 0.0347 | 36 |
| 37 | 1.5835 | 0.6315 | 46.6794 | 0.0214 | 29.4788 | 0.0339 | 37 |
| 38 | 1.6033 | 0.6237 | 48.2629 | 0.0207 | 30.1025 | 0.0332 | 38 |
| 39 | 1.6233 | 0.6160 | 49.8662 | 0.0201 | 30.7185 | 0.0326 | 39 |
| 40 | 1.6436 | 0.6084 | 51.4896 | 0.0194 | 31.3269 | 0.0319 | 40 |
| 50 | 1.8610 | 0.5373 | 68.8818 | 0.0145 | 37.0129 | 0.0270 | 50 |
| 60 | 2.1072 | 0.4746 | 88.5745 | 0.0113 | 42.0346 | 0.0238 | 60 |
| 70 | 2.3859 | 0.4191 | 110.8720 | 0.0090 | 46.4697 | 0.0215 | 70 |
| 80 | 2.7015 | 0.3702 | 136.1188 | 0.0073 | 50.3867 | 0.0198 | 80 |
| 90 | 3.0588 | 0.3269 | 164.7050 | 0.0061 | 53.8461 | 0.0186 | 90 |
| 100 | 3.4634 | 0.2887 | 197.0723 | 0.0051 | 56.9013 | 0.0176 | 100 |
| 110 | 3.9215 | 0.2550 | 233.7209 | 0.0043 | 59.5997 | 0.0168 | 110 |
| 120 | 4.4402 | 0.2252 | 275.2171 | 0.0036 | 61.9828 | 0.0161 | 120 |
| 180 | 9.3563 | 0.1069 | 668.5068 | 0.0015 | 71.4496 | 0.0140 | 180 |
| 240 | 19.7155 | 0.0507 | 1497.2395 | 0.0007 | 75.9423 | 0.0132 | 240 |
| 300 | 41.5441 | 0.0241 | 3243.5296 | 0.0003 | 78.0743 | 0.0128 | 300 |
| 360 | 87.5410 | 0.0114 | 6923.2796 | 0.0001 | 79.0361 | 0.0126 | 360 |

1.50% COMPOUND INTEREST FACTORS

| N | F/P | P/F | F/A | A/F | P/A | A/P | N |
|---|---|---|---|---|---|---|---|
| 1 | 1.0150 | 0.9852 | 1.0000 | 1.0000 | 0.9852 | 1.0150 | 1 |
| 2 | 1.0302 | 0.9707 | 2.0150 | 0.4963 | 1.9559 | 0.5113 | 2 |
| 3 | 1.0457 | 0.9563 | 3.0452 | 0.3284 | 2.9122 | 0.3434 | 3 |
| 4 | 1.0614 | 0.9422 | 4.0909 | 0.2444 | 3.8544 | 0.2594 | 4 |
| 5 | 1.0773 | 0.9283 | 5.1523 | 0.1941 | 4.7826 | 0.2091 | 5 |
| 6 | 1.0934 | 0.9145 | 6.2296 | 0.1605 | 5.6972 | 0.1755 | 6 |
| 7 | 1.1098 | 0.9010 | 7.3230 | 0.1366 | 6.5982 | 0.1516 | 7 |
| 8 | 1.1265 | 0.8877 | 8.4328 | 0.1186 | 7.4859 | 0.1336 | 8 |
| 9 | 1.1434 | 0.8746 | 9.5593 | 0.1046 | 8.3605 | 0.1196 | 9 |
| 10 | 1.1605 | 0.8617 | 10.7027 | 0.0934 | 9.2222 | 0.1084 | 10 |
| 11 | 1.1779 | 0.8489 | 11.8633 | 0.0843 | 10.0711 | 0.0993 | 11 |
| 12 | 1.1956 | 0.8364 | 13.0412 | 0.0767 | 10.9075 | 0.0917 | 12 |
| 13 | 1.2136 | 0.8240 | 14.2368 | 0.0702 | 11.7315 | 0.0852 | 13 |
| 14 | 1.2318 | 0.8118 | 15.4504 | 0.0647 | 12.5434 | 0.0797 | 14 |
| 15 | 1.2502 | 0.7999 | 16.6821 | 0.0599 | 13.3432 | 0.0749 | 15 |
| 16 | 1.2690 | 0.7880 | 17.9324 | 0.0558 | 14.1313 | 0.0708 | 16 |
| 17 | 1.2880 | 0.7764 | 19.2014 | 0.0521 | 14.9076 | 0.0671 | 17 |
| 18 | 1.3073 | 0.7649 | 20.4894 | 0.0488 | 15.6726 | 0.0638 | 18 |
| 19 | 1.3270 | 0.7536 | 21.7967 | 0.0459 | 16.4262 | 0.0609 | 19 |
| 20 | 1.3469 | 0.7425 | 23.1237 | 0.0432 | 17.1686 | 0.0582 | 20 |
| 21 | 1.3671 | 0.7315 | 24.4705 | 0.0409 | 17.9001 | 0.0559 | 21 |
| 22 | 1.3876 | 0.7207 | 25.8376 | 0.0387 | 18.6208 | 0.0537 | 22 |
| 23 | 1.4084 | 0.7100 | 27.2251 | 0.0367 | 19.3309 | 0.0517 | 23 |
| 24 | 1.4295 | 0.6995 | 28.6335 | 0.0349 | 20.0304 | 0.0499 | 24 |
| 25 | 1.4509 | 0.6892 | 30.0630 | 0.0333 | 20.7196 | 0.0483 | 25 |
| 26 | 1.4727 | 0.6790 | 31.5140 | 0.0317 | 21.3986 | 0.0467 | 26 |
| 27 | 1.4948 | 0.6690 | 32.9867 | 0.0303 | 22.0676 | 0.0453 | 27 |
| 28 | 1.5172 | 0.6591 | 34.4815 | 0.0290 | 22.7267 | 0.0440 | 28 |
| 29 | 1.5400 | 0.6494 | 35.9987 | 0.0278 | 23.3761 | 0.0428 | 29 |
| 30 | 1.5631 | 0.6398 | 37.5387 | 0.0266 | 24.0158 | 0.0416 | 30 |
| 31 | 1.5865 | 0.6303 | 39.1018 | 0.0256 | 24.6461 | 0.0406 | 31 |
| 32 | 1.6103 | 0.6210 | 40.6883 | 0.0246 | 25.2671 | 0.0396 | 32 |
| 33 | 1.6345 | 0.6118 | 42.2986 | 0.0236 | 25.8790 | 0.0386 | 33 |
| 34 | 1.6590 | 0.6028 | 43.9331 | 0.0228 | 26.4817 | 0.0378 | 34 |
| 35 | 1.6839 | 0.5939 | 45.5921 | 0.0219 | 27.0756 | 0.0369 | 35 |
| 36 | 1.7091 | 0.5851 | 47.2760 | 0.0212 | 27.6607 | 0.0362 | 36 |
| 37 | 1.7348 | 0.5764 | 48.9851 | 0.0204 | 28.2371 | 0.0354 | 37 |
| 38 | 1.7608 | 0.5679 | 50.7199 | 0.0197 | 28.8051 | 0.0347 | 38 |
| 39 | 1.7872 | 0.5595 | 52.4807 | 0.0191 | 29.3646 | 0.0341 | 39 |
| 40 | 1.8140 | 0.5513 | 54.2679 | 0.0184 | 29.9158 | 0.0334 | 40 |
| 50 | 2.1052 | 0.4750 | 73.6828 | 0.0136 | 34.9997 | 0.0286 | 50 |
| 60 | 2.4432 | 0.4093 | 96.2147 | 0.0104 | 39.3803 | 0.0254 | 60 |
| 70 | 2.8355 | 0.3527 | 122.3638 | 0.0082 | 43.1549 | 0.0232 | 70 |
| 80 | 3.2907 | 0.3039 | 152.7109 | 0.0065 | 46.4073 | 0.0215 | 80 |
| 90 | 3.8189 | 0.2619 | 187.9299 | 0.0053 | 49.2099 | 0.0203 | 90 |
| 100 | 4.4320 | 0.2256 | 228.8030 | 0.0044 | 51.6247 | 0.0194 | 100 |
| 110 | 5.1436 | 0.1944 | 276.2380 | 0.0036 | 53.7055 | 0.0186 | 110 |
| 120 | 5.9693 | 0.1675 | 331.2882 | 0.0030 | 55.4985 | 0.0180 | 120 |
| 180 | 14.5844 | 0.0686 | 905.6245 | 0.0011 | 62.0956 | 0.0161 | 180 |
| 240 | 35.6328 | 0.0281 | 2308.8544 | 0.0004 | 64.7957 | 0.0154 | 240 |
| 300 | 87.0588 | 0.0115 | 5737.2533 | 0.0002 | 65.9009 | 0.0152 | 300 |
| 360 | 212.7038 | 0.0047 | 14113.5854 | 0.0001 | 66.3532 | 0.0151 | 360 |

1.75% COMPOUND INTEREST FACTORS

| N | F/P | P/F | F/A | A/F | P/A | A/P | N |
|---|---|---|---|---|---|---|---|
| 1 | 1.0175 | 0.9828 | 1.0000 | 1.0000 | 0.9828 | 1.0175 | 1 |
| 2 | 1.0353 | 0.9659 | 2.0175 | 0.4957 | 1.9487 | 0.5132 | 2 |
| 3 | 1.0534 | 0.9493 | 3.0528 | 0.3276 | 2.8980 | 0.3451 | 3 |
| 4 | 1.0719 | 0.9330 | 4.1062 | 0.2435 | 3.8309 | 0.2610 | 4 |
| 5 | 1.0906 | 0.9169 | 5.1781 | 0.1931 | 4.7479 | 0.2106 | 5 |
| 6 | 1.1097 | 0.9011 | 6.2687 | 0.1595 | 5.6490 | 0.1770 | 6 |
| 7 | 1.1291 | 0.8856 | 7.3784 | 0.1355 | 6.5346 | 0.1530 | 7 |
| 8 | 1.1489 | 0.8704 | 8.5075 | 0.1175 | 7.4051 | 0.1350 | 8 |
| 9 | 1.1690 | 0.8554 | 9.6564 | 0.1036 | 8.2605 | 0.1211 | 9 |
| 10 | 1.1894 | 0.8407 | 10.8254 | 0.0924 | 9.1012 | 0.1099 | 10 |
| 11 | 1.2103 | 0.8263 | 12.0148 | 0.0832 | 9.9275 | 0.1007 | 11 |
| 12 | 1.2314 | 0.8121 | 13.2251 | 0.0756 | 10.7395 | 0.0931 | 12 |
| 13 | 1.2530 | 0.7981 | 14.4565 | 0.0692 | 11.5376 | 0.0867 | 13 |
| 14 | 1.2749 | 0.7844 | 15.7095 | 0.0637 | 12.3220 | 0.0812 | 14 |
| 15 | 1.2972 | 0.7709 | 16.9844 | 0.0589 | 13.0929 | 0.0764 | 15 |
| 16 | 1.3199 | 0.7576 | 18.2817 | 0.0547 | 13.8505 | 0.0722 | 16 |
| 17 | 1.3430 | 0.7446 | 19.6016 | 0.0510 | 14.5951 | 0.0685 | 17 |
| 18 | 1.3665 | 0.7318 | 20.9446 | 0.0477 | 15.3269 | 0.0652 | 18 |
| 19 | 1.3904 | 0.7192 | 22.3112 | 0.0448 | 16.0461 | 0.0623 | 19 |
| 20 | 1.4148 | 0.7068 | 23.7016 | 0.0422 | 16.7529 | 0.0597 | 20 |
| 21 | 1.4395 | 0.6947 | 25.1164 | 0.0398 | 17.4475 | 0.0573 | 21 |
| 22 | 1.4647 | 0.6827 | 26.5559 | 0.0377 | 18.1303 | 0.0552 | 22 |
| 23 | 1.4904 | 0.6710 | 28.0207 | 0.0357 | 18.8012 | 0.0532 | 23 |
| 24 | 1.5164 | 0.6594 | 29.5110 | 0.0339 | 19.4607 | 0.0514 | 24 |
| 25 | 1.5430 | 0.6481 | 31.0275 | 0.0322 | 20.1088 | 0.0497 | 25 |
| 26 | 1.5700 | 0.6369 | 32.5704 | 0.0307 | 20.7457 | 0.0482 | 26 |
| 27 | 1.5975 | 0.6260 | 34.1404 | 0.0293 | 21.3717 | 0.0468 | 27 |
| 28 | 1.6254 | 0.6152 | 35.7379 | 0.0280 | 21.9870 | 0.0455 | 28 |
| 29 | 1.6539 | 0.6046 | 37.3633 | 0.0268 | 22.5916 | 0.0443 | 29 |
| 30 | 1.6828 | 0.5942 | 39.0172 | 0.0256 | 23.1858 | 0.0431 | 30 |
| 31 | 1.7122 | 0.5840 | 40.7000 | 0.0246 | 23.7699 | 0.0421 | 31 |
| 32 | 1.7422 | 0.5740 | 42.4122 | 0.0236 | 24.3439 | 0.0411 | 32 |
| 33 | 1.7727 | 0.5641 | 44.1544 | 0.0226 | 24.9080 | 0.0401 | 33 |
| 34 | 1.8037 | 0.5544 | 45.9271 | 0.0218 | 25.4624 | 0.0393 | 34 |
| 35 | 1.8353 | 0.5449 | 47.7308 | 0.0210 | 26.0073 | 0.0385 | 35 |
| 36 | 1.8674 | 0.5355 | 49.5661 | 0.0202 | 26.5428 | 0.0377 | 36 |
| 37 | 1.9001 | 0.5263 | 51.4335 | 0.0194 | 27.0690 | 0.0369 | 37 |
| 38 | 1.9333 | 0.5172 | 53.3336 | 0.0187 | 27.5863 | 0.0362 | 38 |
| 39 | 1.9672 | 0.5083 | 55.2670 | 0.0181 | 28.0946 | 0.0356 | 39 |
| 40 | 2.0016 | 0.4996 | 57.2341 | 0.0175 | 28.5942 | 0.0350 | 40 |
| 50 | 2.3808 | 0.4200 | 78.9022 | 0.0127 | 33.1412 | 0.0302 | 50 |
| 60 | 2.8318 | 0.3531 | 104.6752 | 0.0096 | 36.9640 | 0.0271 | 60 |
| 70 | 3.3683 | 0.2969 | 135.3308 | 0.0074 | 40.1779 | 0.0249 | 70 |
| 80 | 4.0064 | 0.2496 | 171.7938 | 0.0058 | 42.8799 | 0.0233 | 80 |
| 90 | 4.7654 | 0.2098 | 215.1646 | 0.0046 | 45.1516 | 0.0221 | 90 |
| 100 | 5.6682 | 0.1764 | 266.7518 | 0.0037 | 47.0615 | 0.0212 | 100 |
| 110 | 6.7420 | 0.1483 | 328.1118 | 0.0030 | 48.6671 | 0.0205 | 110 |
| 120 | 8.0192 | 0.1247 | 401.0962 | 0.0025 | 50.0171 | 0.0200 | 120 |
| 180 | 22.7089 | 0.0440 | 1240.5060 | 0.0008 | 54.6265 | 0.0183 | 180 |
| 240 | 64.3073 | 0.0156 | 3617.5602 | 0.0003 | 56.2543 | 0.0178 | 240 |
| 300 | 182.1065 | 0.0055 | 10348.9410 | 0.0001 | 56.8291 | 0.0176 | 300 |
| 360 | 515.6921 | 0.0019 | 29410.9747 | 0.0000 | 57.0320 | 0.0175 | 360 |

2.00% COMPOUND INTEREST FACTORS

| N | F/P | P/F | F/A | A/F | P/A | A/P | N |
|---|---|---|---|---|---|---|---|
| 1 | 1.0200 | 0.9804 | 1.0000 | 1.0000 | 0.9804 | 1.0200 | 1 |
| 2 | 1.0404 | 0.9612 | 2.0200 | 0.4950 | 1.9416 | 0.5150 | 2 |
| 3 | 1.0612 | 0.9423 | 3.0604 | 0.3268 | 2.8839 | 0.3468 | 3 |
| 4 | 1.0824 | 0.9238 | 4.1216 | 0.2426 | 3.8077 | 0.2626 | 4 |
| 5 | 1.1041 | 0.9057 | 5.2040 | 0.1922 | 4.7135 | 0.2122 | 5 |
| 6 | 1.1262 | 0.8880 | 6.3081 | 0.1585 | 5.6014 | 0.1785 | 6 |
| 7 | 1.1487 | 0.8706 | 7.4343 | 0.1345 | 6.4720 | 0.1545 | 7 |
| 8 | 1.1717 | 0.8535 | 8.5830 | 0.1165 | 7.3255 | 0.1365 | 8 |
| 9 | 1.1951 | 0.8368 | 9.7546 | 0.1025 | 8.1622 | 0.1225 | 9 |
| 10 | 1.2190 | 0.8203 | 10.9497 | 0.0913 | 8.9826 | 0.1113 | 10 |
| 11 | 1.2434 | 0.8043 | 12.1687 | 0.0822 | 9.7868 | 0.1022 | 11 |
| 12 | 1.2682 | 0.7885 | 13.4121 | 0.0746 | 10.5753 | 0.0946 | 12 |
| 13 | 1.2936 | 0.7730 | 14.6803 | 0.0681 | 11.3484 | 0.0881 | 13 |
| 14 | 1.3195 | 0.7579 | 15.9739 | 0.0626 | 12.1062 | 0.0826 | 14 |
| 15 | 1.3459 | 0.7430 | 17.2934 | 0.0578 | 12.8493 | 0.0778 | 15 |
| 16 | 1.3728 | 0.7284 | 18.6393 | 0.0537 | 13.5777 | 0.0737 | 16 |
| 17 | 1.4002 | 0.7142 | 20.0121 | 0.0500 | 14.2919 | 0.0700 | 17 |
| 18 | 1.4282 | 0.7002 | 21.4123 | 0.0467 | 14.9920 | 0.0667 | 18 |
| 19 | 1.4568 | 0.6864 | 22.8406 | 0.0438 | 15.6785 | 0.0638 | 19 |
| 20 | 1.4859 | 0.6730 | 24.2974 | 0.0412 | 16.3514 | 0.0612 | 20 |
| 21 | 1.5157 | 0.6598 | 25.7833 | 0.0388 | 17.0112 | 0.0588 | 21 |
| 22 | 1.5460 | 0.6468 | 27.2990 | 0.0366 | 17.6580 | 0.0566 | 22 |
| 23 | 1.5769 | 0.6342 | 28.8450 | 0.0347 | 18.2922 | 0.0547 | 23 |
| 24 | 1.6084 | 0.6217 | 30.4219 | 0.0329 | 18.9139 | 0.0529 | 24 |
| 25 | 1.6406 | 0.6095 | 32.0303 | 0.0312 | 19.5235 | 0.0512 | 25 |
| 26 | 1.6734 | 0.5976 | 33.6709 | 0.0297 | 20.1210 | 0.0497 | 26 |
| 27 | 1.7069 | 0.5859 | 35.3443 | 0.0283 | 20.7069 | 0.0483 | 27 |
| 28 | 1.7410 | 0.5744 | 37.0512 | 0.0270 | 21.2813 | 0.0470 | 28 |
| 29 | 1.7758 | 0.5631 | 38.7922 | 0.0258 | 21.8444 | 0.0458 | 29 |
| 30 | 1.8114 | 0.5521 | 40.5681 | 0.0246 | 22.3965 | 0.0446 | 30 |
| 31 | 1.8476 | 0.5412 | 42.3794 | 0.0236 | 22.9377 | 0.0436 | 31 |
| 32 | 1.8845 | 0.5306 | 44.2270 | 0.0226 | 23.4683 | 0.0426 | 32 |
| 33 | 1.9222 | 0.5202 | 46.1116 | 0.0217 | 23.9886 | 0.0417 | 33 |
| 34 | 1.9607 | 0.5100 | 48.0338 | 0.0208 | 24.4986 | 0.0408 | 34 |
| 35 | 1.9999 | 0.5000 | 49.9945 | 0.0200 | 24.9986 | 0.0400 | 35 |
| 36 | 2.0399 | 0.4902 | 51.9944 | 0.0192 | 25.4888 | 0.0392 | 36 |
| 37 | 2.0807 | 0.4806 | 54.0343 | 0.0185 | 25.9695 | 0.0385 | 37 |
| 38 | 2.1223 | 0.4712 | 56.1149 | 0.0178 | 26.4406 | 0.0378 | 38 |
| 39 | 2.1647 | 0.4619 | 58.2372 | 0.0172 | 26.9026 | 0.0372 | 39 |
| 40 | 2.2080 | 0.4529 | 60.4020 | 0.0166 | 27.3555 | 0.0366 | 40 |
| 50 | 2.6916 | 0.3715 | 84.5794 | 0.0118 | 31.4236 | 0.0318 | 50 |
| 60 | 3.2810 | 0.3048 | 114.0515 | 0.0088 | 34.7609 | 0.0288 | 60 |
| 70 | 3.9996 | 0.2500 | 149.9779 | 0.0067 | 37.4986 | 0.0267 | 70 |
| 80 | 4.8754 | 0.2051 | 193.7720 | 0.0052 | 39.7445 | 0.0252 | 80 |
| 90 | 5.9431 | 0.1683 | 247.1567 | 0.0040 | 41.5869 | 0.0240 | 90 |
| 100 | 7.2446 | 0.1380 | 312.2323 | 0.0032 | 43.0984 | 0.0232 | 100 |
| 110 | 8.8312 | 0.1132 | 391.5592 | 0.0026 | 44.3382 | 0.0226 | 110 |
| 120 | 10.7652 | 0.0929 | 488.2582 | 0.0020 | 45.3554 | 0.0220 | 120 |
| 180 | 35.3208 | 0.0283 | 1716.0416 | 0.0006 | 48.5844 | 0.0206 | 180 |
| 240 | 115.8887 | 0.0086 | 5744.4368 | 0.0002 | 49.5686 | 0.0202 | 240 |
| 300 | 380.2345 | 0.0026 | 18961.7254 | 0.0001 | 49.8685 | 0.0201 | 300 |
| 360 | 1247.5611 | 0.0008 | 62328.0564 | 0.0000 | 49.9599 | 0.0200 | 360 |

2.50% COMPOUND INTEREST FACTORS

| N | F/P | P/F | F/A | A/F | P/A | A/P | N |
|---|---|---|---|---|---|---|---|
| 1 | 1.0250 | 0.9756 | 1.0000 | 1.0000 | 0.9756 | 1.0250 | 1 |
| 2 | 1.0506 | 0.9518 | 2.0250 | 0.4938 | 1.9274 | 0.5188 | 2 |
| 3 | 1.0769 | 0.9286 | 3.0756 | 0.3251 | 2.8560 | 0.3501 | 3 |
| 4 | 1.1038 | 0.9060 | 4.1525 | 0.2408 | 3.7620 | 0.2658 | 4 |
| 5 | 1.1314 | 0.8839 | 5.2563 | 0.1902 | 4.6458 | 0.2152 | 5 |
| 6 | 1.1597 | 0.8623 | 6.3877 | 0.1565 | 5.5081 | 0.1815 | 6 |
| 7 | 1.1887 | 0.8413 | 7.5474 | 0.1325 | 6.3494 | 0.1575 | 7 |
| 8 | 1.2184 | 0.8207 | 8.7361 | 0.1145 | 7.1701 | 0.1395 | 8 |
| 9 | 1.2489 | 0.8007 | 9.9545 | 0.1005 | 7.9709 | 0.1255 | 9 |
| 10 | 1.2801 | 0.7812 | 11.2034 | 0.0893 | 8.7521 | 0.1143 | 10 |
| 11 | 1.3121 | 0.7621 | 12.4835 | 0.0801 | 9.5142 | 0.1051 | 11 |
| 12 | 1.3449 | 0.7436 | 13.7956 | 0.0725 | 10.2578 | 0.0975 | 12 |
| 13 | 1.3785 | 0.7254 | 15.1404 | 0.0660 | 10.9832 | 0.0910 | 13 |
| 14 | 1.4130 | 0.7077 | 16.5190 | 0.0605 | 11.6909 | 0.0855 | 14 |
| 15 | 1.4483 | 0.6905 | 17.9319 | 0.0558 | 12.3814 | 0.0808 | 15 |
| 16 | 1.4845 | 0.6736 | 19.3802 | 0.0516 | 13.0550 | 0.0766 | 16 |
| 17 | 1.5216 | 0.6572 | 20.8647 | 0.0479 | 13.7122 | 0.0729 | 17 |
| 18 | 1.5597 | 0.6412 | 22.3863 | 0.0447 | 14.3534 | 0.0697 | 18 |
| 19 | 1.5987 | 0.6255 | 23.9460 | 0.0418 | 14.9789 | 0.0668 | 19 |
| 20 | 1.6386 | 0.6103 | 25.5447 | 0.0391 | 15.5892 | 0.0641 | 20 |
| 21 | 1.6796 | 0.5954 | 27.1833 | 0.0368 | 16.1845 | 0.0618 | 21 |
| 22 | 1.7216 | 0.5809 | 28.8629 | 0.0346 | 16.7654 | 0.0596 | 22 |
| 23 | 1.7646 | 0.5667 | 30.5844 | 0.0327 | 17.3321 | 0.0577 | 23 |
| 24 | 1.8087 | 0.5529 | 32.3490 | 0.0309 | 17.8850 | 0.0559 | 24 |
| 25 | 1.8539 | 0.5394 | 34.1578 | 0.0293 | 18.4244 | 0.0543 | 25 |
| 26 | 1.9003 | 0.5262 | 36.0117 | 0.0278 | 18.9506 | 0.0528 | 26 |
| 27 | 1.9478 | 0.5134 | 37.9120 | 0.0264 | 19.4640 | 0.0514 | 27 |
| 28 | 1.9965 | 0.5009 | 39.8598 | 0.0251 | 19.9649 | 0.0501 | 28 |
| 29 | 2.0464 | 0.4887 | 41.8563 | 0.0239 | 20.4535 | 0.0489 | 29 |
| 30 | 2.0976 | 0.4767 | 43.9027 | 0.0228 | 20.9303 | 0.0478 | 30 |
| 31 | 2.1500 | 0.4651 | 46.0003 | 0.0217 | 21.3954 | 0.0467 | 31 |
| 32 | 2.2038 | 0.4538 | 48.1503 | 0.0208 | 21.8492 | 0.0458 | 32 |
| 33 | 2.2589 | 0.4427 | 50.3540 | 0.0199 | 22.2919 | 0.0449 | 33 |
| 34 | 2.3153 | 0.4319 | 52.6129 | 0.0190 | 22.7238 | 0.0440 | 34 |
| 35 | 2.3732 | 0.4214 | 54.9282 | 0.0182 | 23.1452 | 0.0432 | 35 |
| 36 | 2.4325 | 0.4111 | 57.3014 | 0.0175 | 23.5563 | 0.0425 | 36 |
| 37 | 2.4933 | 0.4011 | 59.7339 | 0.0167 | 23.9573 | 0.0417 | 37 |
| 38 | 2.5557 | 0.3913 | 62.2273 | 0.0161 | 24.3486 | 0.0411 | 38 |
| 39 | 2.6196 | 0.3817 | 64.7830 | 0.0154 | 24.7303 | 0.0404 | 39 |
| 40 | 2.6851 | 0.3724 | 67.4026 | 0.0148 | 25.1028 | 0.0398 | 40 |
| 50 | 3.4371 | 0.2909 | 97.4843 | 0.0103 | 28.3623 | 0.0353 | 50 |
| 60 | 4.3998 | 0.2273 | 135.9916 | 0.0074 | 30.9087 | 0.0324 | 60 |
| 70 | 5.6321 | 0.1776 | 185.2841 | 0.0054 | 32.8979 | 0.0304 | 70 |
| 80 | 7.2096 | 0.1387 | 248.3827 | 0.0040 | 34.4518 | 0.0290 | 80 |
| 90 | 9.2289 | 0.1084 | 329.1543 | 0.0030 | 35.6658 | 0.0280 | 90 |
| 100 | 11.8137 | 0.0846 | 432.5487 | 0.0023 | 36.6141 | 0.0273 | 100 |
| 110 | 15.1226 | 0.0661 | 564.9022 | 0.0018 | 37.3549 | 0.0268 | 110 |
| 120 | 19.3581 | 0.0517 | 734.3260 | 0.0014 | 37.9337 | 0.0264 | 120 |
| 180 | 85.1718 | 0.0117 | 3366.8716 | 0.0003 | 39.5304 | 0.0253 | 180 |
| 240 | 374.7380 | 0.0027 | 14949.5186 | 0.0001 | 39.8933 | 0.0251 | 240 |
| 300 | 1648.7683 | 0.0006 | 65910.7303 | 0.0000 * | 39.9757 | 0.0250 | 300 |

3.00% COMPOUND INTEREST FACTORS

| N | F/P | P/F | F/A | A/F | P/A | A/P | N |
|---|---|---|---|---|---|---|---|
| 1 | 1.0300 | 0.9709 | 1.0000 | 1.0000 | 0.9709 | 1.0300 | 1 |
| 2 | 1.0609 | 0.9426 | 2.0300 | 0.4926 | 1.9135 | 0.5226 | 2 |
| 3 | 1.0927 | 0.9151 | 3.0909 | 0.3235 | 2.8286 | 0.3535 | 3 |
| 4 | 1.1255 | 0.8885 | 4.1836 | 0.2390 | 3.7171 | 0.2690 | 4 |
| 5 | 1.1593 | 0.8626 | 5.3091 | 0.1884 | 4.5797 | 0.2184 | 5 |
| 6 | 1.1941 | 0.8375 | 6.4684 | 0.1546 | 5.4172 | 0.1846 | 6 |
| 7 | 1.2299 | 0.8131 | 7.6625 | 0.1305 | 6.2303 | 0.1605 | 7 |
| 8 | 1.2668 | 0.7894 | 8.8923 | 0.1125 | 7.0197 | 0.1425 | 8 |
| 9 | 1.3048 | 0.7664 | 10.1591 | 0.0984 | 7.7861 | 0.1284 | 9 |
| 10 | 1.3439 | 0.7441 | 11.4639 | 0.0872 | 8.5302 | 0.1172 | 10 |
| 11 | 1.3842 | 0.7224 | 12.8078 | 0.0781 | 9.2526 | 0.1081 | 11 |
| 12 | 1.4258 | 0.7014 | 14.1920 | 0.0705 | 9.9540 | 0.1005 | 12 |
| 13 | 1.4685 | 0.6810 | 15.6178 | 0.0640 | 10.6350 | 0.0940 | 13 |
| 14 | 1.5126 | 0.6611 | 17.0863 | 0.0585 | 11.2961 | 0.0885 | 14 |
| 15 | 1.5580 | 0.6419 | 18.5989 | 0.0538 | 11.9379 | 0.0838 | 15 |
| 16 | 1.6047 | 0.6232 | 20.1569 | 0.0496 | 12.5611 | 0.0796 | 16 |
| 17 | 1.6528 | 0.6050 | 21.7616 | 0.0460 | 13.1661 | 0.0760 | 17 |
| 18 | 1.7024 | 0.5874 | 23.4144 | 0.0427 | 13.7535 | 0.0727 | 18 |
| 19 | 1.7535 | 0.5703 | 25.1169 | 0.0398 | 14.3238 | 0.0698 | 19 |
| 20 | 1.8061 | 0.5537 | 26.8704 | 0.0372 | 14.8775 | 0.0672 | 20 |
| 21 | 1.8603 | 0.5375 | 28.6765 | 0.0349 | 15.4150 | 0.0649 | 21 |
| 22 | 1.9161 | 0.5219 | 30.5368 | 0.0327 | 15.9369 | 0.0627 | 22 |
| 23 | 1.9736 | 0.5067 | 32.4529 | 0.0308 | 16.4436 | 0.0608 | 23 |
| 24 | 2.0328 | 0.4919 | 34.4265 | 0.0290 | 16.9355 | 0.0590 | 24 |
| 25 | 2.0938 | 0.4776 | 36.4593 | 0.0274 | 17.4131 | 0.0574 | 25 |
| 26 | 2.1566 | 0.4637 | 38.5530 | 0.0259 | 17.8768 | 0.0559 | 26 |
| 27 | 2.2213 | 0.4502 | 40.7096 | 0.0246 | 18.3270 | 0.0546 | 27 |
| 28 | 2.2879 | 0.4371 | 42.9309 | 0.0233 | 18.7641 | 0.0533 | 28 |
| 29 | 2.3566 | 0.4243 | 45.2189 | 0.0221 | 19.1885 | 0.0521 | 29 |
| 30 | 2.4273 | 0.4120 | 47.5754 | 0.0210 | 19.6004 | 0.0510 | 30 |
| 31 | 2.5001 | 0.4000 | 50.0027 | 0.0200 | 20.0004 | 0.0500 | 31 |
| 32 | 2.5751 | 0.3883 | 52.5028 | 0.0190 | 20.3888 | 0.0490 | 32 |
| 33 | 2.6523 | 0.3770 | 55.0778 | 0.0182 | 20.7658 | 0.0482 | 33 |
| 34 | 2.7319 | 0.3660 | 57.7302 | 0.0173 | 21.1318 | 0.0473 | 34 |
| 35 | 2.8139 | 0.3554 | 60.4621 | 0.0165 | 21.4872 | 0.0465 | 35 |
| 36 | 2.8983 | 0.3450 | 63.2759 | 0.0158 | 21.8323 | 0.0458 | 36 |
| 37 | 2.9852 | 0.3350 | 66.1742 | 0.0151 | 22.1672 | 0.0451 | 37 |
| 38 | 3.0748 | 0.3252 | 69.1594 | 0.0145 | 22.4925 | 0.0445 | 38 |
| 39 | 3.1670 | 0.3158 | 72.2342 | 0.0138 | 22.8082 | 0.0438 | 39 |
| 40 | 3.2620 | 0.3066 | 75.4013 | 0.0133 | 23.1148 | 0.0433 | 40 |
| 50 | 4.3839 | 0.2281 | 112.7969 | 0.0089 | 25.7298 | 0.0389 | 50 |
| 60 | 5.8916 | 0.1697 | 163.0534 | 0.0061 | 27.6756 | 0.0361 | 60 |
| 70 | 7.9178 | 0.1263 | 230.5941 | 0.0043 | 29.1234 | 0.0343 | 70 |
| 80 | 10.6409 | 0.0940 | 321.3630 | 0.0031 | 30.2008 | 0.0331 | 80 |
| 90 | 14.3005 | 0.0699 | 443.3489 | 0.0023 | 31.0024 | 0.0323 | 90 |
| 100 | 19.2186 | 0.0520 | 607.2877 | 0.0016 | 31.5989 | 0.0316 | 100 |
| 110 | 25.8282 | 0.0387 | 827.6078 | 0.0012 | 32.0428 | 0.0312 | 110 |
| 120 | 34.7110 | 0.0288 | 1123.6996 | 0.0009 | 32.3730 | 0.0309 | 120 |
| 180 | 204.5034 | 0.0049 | 6783.4453 | 0.0001 | 33.1703 | 0.0301 | 180 |
| 240 | 1204.8526 | 0.0008 | 40128.4209 | 0.0000 | 33.3057 | 0.0300 | 240 |

4.00% COMPOUND INTEREST FACTORS

| N | F/P | P/F | F/A | A/F | P/A | A/P | N |
|---|---|---|---|---|---|---|---|
| 1 | 1.0400 | 0.9615 | 1.0000 | 1.0000 | 0.9615 | 1.0400 | 1 |
| 2 | 1.0816 | 0.9246 | 2.0400 | 0.4902 | 1.8861 | 0.5302 | 2 |
| 3 | 1.1249 | 0.8890 | 3.1216 | 0.3203 | 2.7751 | 0.3603 | 3 |
| 4 | 1.1699 | 0.8548 | 4.2465 | 0.2355 | 3.6299 | 0.2755 | 4 |
| 5 | 1.2167 | 0.8219 | 5.4163 | 0.1846 | 4.4518 | 0.2246 | 5 |
| 6 | 1.2653 | 0.7903 | 6.6330 | 0.1508 | 5.2421 | 0.1908 | 6 |
| 7 | 1.3159 | 0.7599 | 7.8983 | 0.1266 | 6.0021 | 0.1666 | 7 |
| 8 | 1.3686 | 0.7307 | 9.2142 | 0.1085 | 6.7327 | 0.1485 | 8 |
| 9 | 1.4233 | 0.7026 | 10.5828 | 0.0945 | 7.4353 | 0.1345 | 9 |
| 10 | 1.4802 | 0.6756 | 12.0061 | 0.0833 | 8.1109 | 0.1233 | 10 |
| 11 | 1.5395 | 0.6496 | 13.4864 | 0.0741 | 8.7605 | 0.1141 | 11 |
| 12 | 1.6010 | 0.6246 | 15.0258 | 0.0666 | 9.3851 | 0.1066 | 12 |
| 13 | 1.6651 | 0.6006 | 16.6268 | 0.0601 | 9.9856 | 0.1001 | 13 |
| 14 | 1.7317 | 0.5775 | 18.2919 | 0.0547 | 10.5631 | 0.0947 | 14 |
| 15 | 1.8009 | 0.5553 | 20.0236 | 0.0499 | 11.1184 | 0.0899 | 15 |
| 16 | 1.8730 | 0.5339 | 21.8245 | 0.0458 | 11.6523 | 0.0858 | 16 |
| 17 | 1.9479 | 0.5134 | 23.6975 | 0.0422 | 12.1657 | 0.0822 | 17 |
| 18 | 2.0258 | 0.4936 | 25.6454 | 0.0390 | 12.6593 | 0.0790 | 18 |
| 19 | 2.1068 | 0.4746 | 27.6712 | 0.0361 | 13.1339 | 0.0761 | 19 |
| 20 | 2.1911 | 0.4564 | 29.7781 | 0.0336 | 13.5903 | 0.0736 | 20 |
| 21 | 2.2788 | 0.4388 | 31.9692 | 0.0313 | 14.0292 | 0.0713 | 21 |
| 22 | 2.3699 | 0.4220 | 34.2480 | 0.0292 | 14.4511 | 0.0692 | 22 |
| 23 | 2.4647 | 0.4057 | 36.6179 | 0.0273 | 14.8568 | 0.0673 | 23 |
| 24 | 2.5633 | 0.3901 | 39.0826 | 0.0256 | 15.2470 | 0.0656 | 24 |
| 25 | 2.6658 | 0.3751 | 41.6459 | 0.0240 | 15.6221 | 0.0640 | 25 |
| 26 | 2.7725 | 0.3607 | 44.3117 | 0.0226 | 15.9828 | 0.0626 | 26 |
| 27 | 2.8834 | 0.3468 | 47.0842 | 0.0212 | 16.3296 | 0.0612 | 27 |
| 28 | 2.9987 | 0.3335 | 49.9676 | 0.0200 | 16.6631 | 0.0600 | 28 |
| 29 | 3.1187 | 0.3207 | 52.9663 | 0.0189 | 16.9837 | 0.0589 | 29 |
| 30 | 3.2434 | 0.3083 | 56.0849 | 0.0178 | 17.2920 | 0.0578 | 30 |
| 31 | 3.3731 | 0.2965 | 59.3283 | 0.0169 | 17.5885 | 0.0569 | 31 |
| 32 | 3.5081 | 0.2851 | 62.7015 | 0.0159 | 17.8736 | 0.0559 | 32 |
| 33 | 3.6484 | 0.2741 | 66.2095 | 0.0151 | 18.1476 | 0.0551 | 33 |
| 34 | 3.7943 | 0.2636 | 69.8579 | 0.0143 | 18.4112 | 0.0543 | 34 |
| 35 | 3.9461 | 0.2534 | 73.6522 | 0.0136 | 18.6646 | 0.0536 | 35 |
| 36 | 4.1039 | 0.2437 | 77.5983 | 0.0129 | 18.9083 | 0.0529 | 36 |
| 37 | 4.2681 | 0.2343 | 81.7022 | 0.0122 | 19.1426 | 0.0522 | 37 |
| 38 | 4.4388 | 0.2253 | 85.9703 | 0.0116 | 19.3679 | 0.0516 | 38 |
| 39 | 4.6164 | 0.2166 | 90.4091 | 0.0111 | 19.5845 | 0.0511 | 39 |
| 40 | 4.8010 | 0.2083 | 95.0255 | 0.0105 | 19.7928 | 0.0505 | 40 |
| 50 | 7.1067 | 0.1407 | 152.6671 | 0.0066 | 21.4822 | 0.0466 | 50 |
| 60 | 10.5196 | 0.0951 | 237.9907 | 0.0042 | 22.6235 | 0.0442 | 60 |
| 70 | 15.5716 | 0.0642 | 364.2905 | 0.0027 | 23.3945 | 0.0427 | 70 |
| 80 | 23.0498 | 0.0434 | 551.2450 | 0.0018 | 23.9154 | 0.0418 | 80 |
| 90 | 34.1193 | 0.0293 | 827.9833 | 0.0012 | 24.2673 | 0.0412 | 90 |
| 100 | 50.5049 | 0.0198 | 1237.6237 | 0.0008 | 24.5050 | 0.0408 | 100 |
| 110 | 74.7597 | 0.0134 | 1843.9915 | 0.0005 | 24.6656 | 0.0405 | 110 |
| 120 | 110.6626 | 0.0090 | 2741.5640 | 0.0004 | 24.7741 | 0.0404 | 120 |
| 180 | 1164.1289 | 0.0009 | 29078.2227 | 0.0000 | 24.9785 | 0.0400 | 180 |

5.00% COMPOUND INTEREST FACTORS

| N | F/P | P/F | F/A | A/F | P/A | A/P | N |
|---|---|---|---|---|---|---|---|
| 1 | 1.0500 | 0.9524 | 1.0000 | 1.0000 | 0.9524 | 1.0500 | 1 |
| 2 | 1.1025 | 0.9070 | 2.0500 | 0.4878 | 1.8594 | 0.5378 | 2 |
| 3 | 1.1576 | 0.8638 | 3.1525 | 0.3172 | 2.7232 | 0.3672 | 3 |
| 4 | 1.2155 | 0.8227 | 4.3101 | 0.2320 | 3.5460 | 0.2820 | 4 |
| 5 | 1.2763 | 0.7835 | 5.5256 | 0.1810 | 4.3295 | 0.2310 | 5 |
| 6 | 1.3401 | 0.7462 | 6.8019 | 0.1470 | 5.0757 | 0.1970 | 6 |
| 7 | 1.4071 | 0.7107 | 8.1420 | 0.1228 | 5.7864 | 0.1728 | 7 |
| 8 | 1.4775 | 0.6768 | 9.5491 | 0.1047 | 6.4632 | 0.1547 | 8 |
| 9 | 1.5513 | 0.6446 | 11.0266 | 0.0907 | 7.1078 | 0.1407 | 9 |
| 10 | 1.6289 | 0.6139 | 12.5779 | 0.0795 | 7.7217 | 0.1295 | 10 |
| 11 | 1.7103 | 0.5847 | 14.2068 | 0.0704 | 8.3064 | 0.1204 | 11 |
| 12 | 1.7959 | 0.5568 | 15.9171 | 0.0628 | 8.8633 | 0.1128 | 12 |
| 13 | 1.8856 | 0.5303 | 17.7130 | 0.0565 | 9.3936 | 0.1065 | 13 |
| 14 | 1.9799 | 0.5051 | 19.5986 | 0.0510 | 9.8986 | 0.1010 | 14 |
| 15 | 2.0789 | 0.4810 | 21.5786 | 0.0463 | 10.3797 | 0.0963 | 15 |
| 16 | 2.1829 | 0.4581 | 23.6575 | 0.0423 | 10.8378 | 0.0923 | 16 |
| 17 | 2.2920 | 0.4363 | 25.8404 | 0.0387 | 11.2741 | 0.0887 | 17 |
| 18 | 2.4066 | 0.4155 | 28.1324 | 0.0355 | 11.6896 | 0.0855 | 18 |
| 19 | 2.5270 | 0.3957 | 30.5390 | 0.0327 | 12.0853 | 0.0827 | 19 |
| 20 | 2.6533 | 0.3769 | 33.0660 | 0.0302 | 12.4622 | 0.0802 | 20 |
| 21 | 2.7860 | 0.3589 | 35.7193 | 0.0280 | 12.8212 | 0.0780 | 21 |
| 22 | 2.9253 | 0.3418 | 38.5052 | 0.0260 | 13.1630 | 0.0760 | 22 |
| 23 | 3.0715 | 0.3256 | 41.4305 | 0.0241 | 13.4886 | 0.0741 | 23 |
| 24 | 3.2251 | 0.3101 | 44.5020 | 0.0225 | 13.7986 | 0.0725 | 24 |
| 25 | 3.3864 | 0.2953 | 47.7271 | 0.0210 | 14.0939 | 0.0710 | 25 |
| 26 | 3.5557 | 0.2812 | 51.1135 | 0.0196 | 14.3752 | 0.0696 | 26 |
| 27 | 3.7335 | 0.2678 | 54.6691 | 0.0183 | 14.6430 | 0.0683 | 27 |
| 28 | 3.9201 | 0.2551 | 58.4026 | 0.0171 | 14.8981 | 0.0671 | 28 |
| 29 | 4.1161 | 0.2429 | 62.3227 | 0.0160 | 15.1411 | 0.0660 | 29 |
| 30 | 4.3219 | 0.2314 | 66.4388 | 0.0151 | 15.3725 | 0.0651 | 30 |
| 31 | 4.5380 | 0.2204 | 70.7608 | 0.0141 | 15.5928 | 0.0641 | 31 |
| 32 | 4.7649 | 0.2099 | 75.2988 | 0.0133 | 15.8027 | 0.0633 | 32 |
| 33 | 5.0032 | 0.1999 | 80.0638 | 0.0125 | 16.0025 | 0.0625 | 33 |
| 34 | 5.2533 | 0.1904 | 85.0670 | 0.0118 | 16.1929 | 0.0618 | 34 |
| 35 | 5.5160 | 0.1813 | 90.3203 | 0.0111 | 16.3742 | 0.0611 | 35 |
| 36 | 5.7918 | 0.1727 | 95.8363 | 0.0104 | 16.5469 | 0.0604 | 36 |
| 37 | 6.0814 | 0.1644 | 101.6281 | 0.0098 | 16.7113 | 0.0598 | 37 |
| 38 | 6.3855 | 0.1566 | 107.7095 | 0.0093 | 16.8679 | 0.0593 | 38 |
| 39 | 6.7048 | 0.1491 | 114.0950 | 0.0088 | 17.0170 | 0.0588 | 39 |
| 40 | 7.0400 | 0.1420 | 120.7998 | 0.0083 | 17.1591 | 0.0583 | 40 |
| 50 | 11.4674 | 0.0872 | 209.3480 | 0.0048 | 18.2559 | 0.0548 | 50 |
| 60 | 18.6792 | 0.0535 | 353.5837 | 0.0028 | 18.9293 | 0.0528 | 60 |
| 70 | 30.4264 | 0.0329 | 588.5285 | 0.0017 | 19.3427 | 0.0517 | 70 |
| 80 | 49.5614 | 0.0202 | 971.2288 | 0.0010 | 19.5965 | 0.0510 | 80 |
| 90 | 80.7304 | 0.0124 | 1594.6073 | 0.0006 | 19.7523 | 0.0506 | 90 |
| 100 | 131.5013 | 0.0076 | 2610.0252 | 0.0004 | 19.8479 | 0.0504 | 100 |
| 110 | 214.2017 | 0.0047 | 4264.0338 | 0.0002 | 19.9066 | 0.0502 | 110 |
| 120 | 348.9120 | 0.0029 | 6958.2397 | 0.0001 | 19.9427 | 0.0501 | 120 |

6.00% COMPOUND INTEREST FACTORS

| N | F/P | P/F | F/A | A/F | P/A | A/P | N |
|---|---|---|---|---|---|---|---|
| 1 | 1.0600 | 0.9434 | 1.0000 | 1.0000 | 0.9434 | 1.0600 | 1 |
| 2 | 1.1236 | 0.8900 | 2.0600 | 0.4854 | 1.8334 | 0.5454 | 2 |
| 3 | 1.1910 | 0.8396 | 3.1836 | 0.3141 | 2.6730 | 0.3741 | 3 |
| 4 | 1.2625 | 0.7921 | 4.3746 | 0.2286 | 3.4651 | 0.2886 | 4 |
| 5 | 1.3382 | 0.7473 | 5.6371 | 0.1774 | 4.2124 | 0.2374 | 5 |
| 6 | 1.4185 | 0.7050 | 6.9753 | 0.1434 | 4.9173 | 0.2034 | 6 |
| 7 | 1.5036 | 0.6651 | 8.3938 | 0.1191 | 5.5824 | 0.1791 | 7 |
| 8 | 1.5938 | 0.6274 | 9.8975 | 0.1010 | 6.2098 | 0.1610 | 8 |
| 9 | 1.6895 | 0.5919 | 11.4913 | 0.0870 | 6.8017 | 0.1470 | 9 |
| 10 | 1.7908 | 0.5584 | 13.1808 | 0.0759 | 7.3601 | 0.1359 | 10 |
| 11 | 1.8983 | 0.5268 | 14.9716 | 0.0668 | 7.8869 | 0.1268 | 11 |
| 12 | 2.0122 | 0.4970 | 16.8699 | 0.0593 | 8.3838 | 0.1193 | 12 |
| 13 | 2.1329 | 0.4688 | 18.8821 | 0.0530 | 8.8527 | 0.1130 | 13 |
| 14 | 2.2609 | 0.4423 | 21.0151 | 0.0476 | 9.2950 | 0.1076 | 14 |
| 15 | 2.3966 | 0.4173 | 23.2760 | 0.0430 | 9.7122 | 0.1030 | 15 |
| 16 | 2.5404 | 0.3936 | 25.6725 | 0.0390 | 10.1059 | 0.0990 | 16 |
| 17 | 2.6928 | 0.3714 | 28.2129 | 0.0354 | 10.4773 | 0.0954 | 17 |
| 18 | 2.8543 | 0.3503 | 30.9057 | 0.0324 | 10.8276 | 0.0924 | 18 |
| 19 | 3.0256 | 0.3305 | 33.7600 | 0.0296 | 11.1581 | 0.0896 | 19 |
| 20 | 3.2071 | 0.3118 | 36.7856 | 0.0272 | 11.4699 | 0.0872 | 20 |
| 21 | 3.3996 | 0.2942 | 39.9927 | 0.0250 | 11.7641 | 0.0850 | 21 |
| 22 | 3.6035 | 0.2775 | 43.3923 | 0.0230 | 12.0416 | 0.0830 | 22 |
| 23 | 3.8197 | 0.2618 | 46.9958 | 0.0213 | 12.3034 | 0.0813 | 23 |
| 24 | 4.0489 | 0.2470 | 50.8156 | 0.0197 | 12.5504 | 0.0797 | 24 |
| 25 | 4.2919 | 0.2330 | 54.8645 | 0.0182 | 12.7834 | 0.0782 | 25 |
| 26 | 4.5494 | 0.2198 | 59.1564 | 0.0169 | 13.0032 | 0.0769 | 26 |
| 27 | 4.8223 | 0.2074 | 63.7058 | 0.0157 | 13.2105 | 0.0757 | 27 |
| 28 | 5.1117 | 0.1956 | 68.5281 | 0.0146 | 13.4062 | 0.0746 | 28 |
| 29 | 5.4184 | 0.1846 | 73.6398 | 0.0136 | 13.5907 | 0.0736 | 29 |
| 30 | 5.7435 | 0.1741 | 79.0582 | 0.0126 | 13.7648 | 0.0726 | 30 |
| 31 | 6.0881 | 0.1643 | 84.8017 | 0.0118 | 13.9291 | 0.0718 | 31 |
| 32 | 6.4534 | 0.1550 | 90.8898 | 0.0110 | 14.0840 | 0.0710 | 32 |
| 33 | 6.8406 | 0.1462 | 97.3432 | 0.0103 | 14.2302 | 0.0703 | 33 |
| 34 | 7.2510 | 0.1379 | 104.1838 | 0.0096 | 14.3681 | 0.0696 | 34 |
| 35 | 7.6861 | 0.1301 | 111.4348 | 0.0090 | 14.4982 | 0.0690 | 35 |
| 36 | 8.1473 | 0.1227 | 119.1209 | 0.0084 | 14.6210 | 0.0684 | 36 |
| 37 | 8.6361 | 0.1158 | 127.2681 | 0.0079 | 14.7368 | 0.0679 | 37 |
| 38 | 9.1543 | 0.1092 | 135.9042 | 0.0074 | 14.8460 | 0.0674 | 38 |
| 39 | 9.7035 | 0.1031 | 145.0585 | 0.0069 | 14.9491 | 0.0669 | 39 |
| 40 | 10.2857 | 0.0972 | 154.7620 | 0.0065 | 15.0463 | 0.0665 | 40 |
| 50 | 18.4202 | 0.0543 | 290.3359 | 0.0034 | 15.7619 | 0.0634 | 50 |
| 60 | 32.9877 | 0.0303 | 533.1282 | 0.0019 | 16.1614 | 0.0619 | 60 |
| 70 | 59.0759 | 0.0169 | 967.9322 | 0.0010 | 16.3845 | 0.0610 | 70 |
| 80 | 105.7960 | 0.0095 | 1746.5999 | 0.0006 | 16.5091 | 0.0606 | 80 |
| 90 | 189.4645 | 0.0053 | 3141.0752 | 0.0003 | 16.5787 | 0.0603 | 90 |
| 100 | 339.3021 | 0.0029 | 5638.3681 | 0.0002 | 16.6175 | 0.0602 | 100 |
| 110 | 607.6384 | 0.0016 | 10110.6392 | 0.0001 | 16.6392 | 0.0601 | 110 |
| 120 | 1088.1877 | 0.0009 | 18119.7958 | 0.0001 | 16.6514 | 0.0601 | 120 |

7.00% COMPOUND INTEREST FACTORS

| N | F/P | P/F | F/A | A/F | P/A | A/P | N |
|---|---|---|---|---|---|---|---|
| 1 | 1.0700 | 0.9346 | 1.0000 | 1.0000 | 0.9346 | 1.0700 | 1 |
| 2 | 1.1449 | 0.8734 | 2.0700 | 0.4831 | 1.8080 | 0.5531 | 2 |
| 3 | 1.2250 | 0.8163 | 3.2149 | 0.3111 | 2.6243 | 0.3811 | 3 |
| 4 | 1.3108 | 0.7629 | 4.4399 | 0.2252 | 3.3872 | 0.2952 | 4 |
| 5 | 1.4026 | 0.7130 | 5.7507 | 0.1739 | 4.1002 | 0.2439 | 5 |
| 6 | 1.5007 | 0.6663 | 7.1533 | 0.1398 | 4.7665 | 0.2098 | 6 |
| 7 | 1.6058 | 0.6227 | 8.6540 | 0.1156 | 5.3893 | 0.1856 | 7 |
| 8 | 1.7182 | 0.5820 | 10.2598 | 0.0975 | 5.9713 | 0.1675 | 8 |
| 9 | 1.8385 | 0.5439 | 11.9780 | 0.0835 | 6.5152 | 0.1535 | 9 |
| 10 | 1.9672 | 0.5083 | 13.8164 | 0.0724 | 7.0236 | 0.1424 | 10 |
| 11 | 2.1049 | 0.4751 | 15.7836 | 0.0634 | 7.4987 | 0.1334 | 11 |
| 12 | 2.2522 | 0.4440 | 17.8885 | 0.0559 | 7.9427 | 0.1259 | 12 |
| 13 | 2.4098 | 0.4150 | 20.1406 | 0.0497 | 8.3577 | 0.1197 | 13 |
| 14 | 2.5785 | 0.3878 | 22.5505 | 0.0443 | 8.7455 | 0.1143 | 14 |
| 15 | 2.7590 | 0.3624 | 25.1290 | 0.0398 | 9.1079 | 0.1098 | 15 |
| 16 | 2.9522 | 0.3387 | 27.8881 | 0.0359 | 9.4466 | 0.1059 | 16 |
| 17 | 3.1588 | 0.3166 | 30.8402 | 0.0324 | 9.7632 | 0.1024 | 17 |
| 18 | 3.3799 | 0.2959 | 33.9990 | 0.0294 | 10.0591 | 0.0994 | 18 |
| 19 | 3.6165 | 0.2765 | 37.3790 | 0.0268 | 10.3356 | 0.0968 | 19 |
| 20 | 3.8697 | 0.2584 | 40.9955 | 0.0244 | 10.5940 | 0.0944 | 20 |
| 21 | 4.1406 | 0.2415 | 44.8652 | 0.0223 | 10.8355 | 0.0923 | 21 |
| 22 | 4.4304 | 0.2257 | 49.0057 | 0.0204 | 11.0612 | 0.0904 | 22 |
| 23 | 4.7405 | 0.2109 | 53.4361 | 0.0187 | 11.2722 | 0.0887 | 23 |
| 24 | 5.0724 | 0.1971 | 58.1767 | 0.0172 | 11.4693 | 0.0872 | 24 |
| 25 | 5.4274 | 0.1842 | 63.2490 | 0.0158 | 11.6536 | 0.0858 | 25 |
| 26 | 5.8074 | 0.1722 | 68.6765 | 0.0146 | 11.8258 | 0.0846 | 26 |
| 27 | 6.2139 | 0.1609 | 74.4838 | 0.0134 | 11.9867 | 0.0834 | 27 |
| 28 | 6.6488 | 0.1504 | 80.6977 | 0.0124 | 12.1371 | 0.0824 | 28 |
| 29 | 7.1143 | 0.1406 | 87.3465 | 0.0114 | 12.2777 | 0.0814 | 29 |
| 30 | 7.6123 | 0.1314 | 94.4608 | 0.0106 | 12.4090 | 0.0806 | 30 |
| 31 | 8.1451 | 0.1228 | 102.0730 | 0.0098 | 12.5318 | 0.0798 | 31 |
| 32 | 8.7153 | 0.1147 | 110.2182 | 0.0091 | 12.6466 | 0.0791 | 32 |
| 33 | 9.3253 | 0.1072 | 118.9334 | 0.0084 | 12.7538 | 0.0784 | 33 |
| 34 | 9.9781 | 0.1002 | 128.2588 | 0.0078 | 12.8540 | 0.0778 | 34 |
| 35 | 10.6766 | 0.0937 | 138.2369 | 0.0072 | 12.9477 | 0.0772 | 35 |
| 36 | 11.4239 | 0.0875 | 148.9135 | 0.0067 | 13.0352 | 0.0767 | 36 |
| 37 | 12.2236 | 0.0818 | 160.3374 | 0.0062 | 13.1170 | 0.0762 | 37 |
| 38 | 13.0793 | 0.0765 | 172.5610 | 0.0058 | 13.1935 | 0.0758 | 38 |
| 39 | 13.9948 | 0.0715 | 185.6403 | 0.0054 | 13.2649 | 0.0754 | 39 |
| 40 | 14.9745 | 0.0668 | 199.6351 | 0.0050 | 13.3317 | 0.0750 | 40 |
| 50 | 29.4570 | 0.0339 | 406.5289 | 0.0025 | 13.8007 | 0.0725 | 50 |
| 60 | 57.9464 | 0.0173 | 813.5204 | 0.0012 | 14.0392 | 0.0712 | 60 |
| 70 | 113.9894 | 0.0088 | 1614.1342 | 0.0006 | 14.1604 | 0.0706 | 70 |
| 80 | 224.2344 | 0.0045 | 3189.0627 | 0.0003 | 14.2220 | 0.0703 | 80 |
| 90 | 441.1030 | 0.0023 | 6287.1854 | 0.0002 | 14.2533 | 0.0702 | 90 |
| 100 | 867.7163 | 0.0012 | 12381.6618 | 0.0001 | 14.2693 | 0.0701 | 100 |
| 110 | 1706.9293 | 0.0006 | 24370.4193 | 0.0000 | 14.2773 | 0.0700 | 110 |
| 120 | 3357.7884 | 0.0003 | 47954.1198 | 0.0000 | 14.2815 | 0.0700 | 120 |

8.00% COMPOUND INTEREST FACTORS

| N | F/P | P/F | F/A | A/F | P/A | A/P | N |
|---|---|---|---|---|---|---|---|
| 1 | 1.0800 | 0.9259 | 1.0000 | 1.0000 | 0.9259 | 1.0800 | 1 |
| 2 | 1.1664 | 0.8573 | 2.0800 | 0.4808 | 1.7833 | 0.5608 | 2 |
| 3 | 1.2597 | 0.7938 | 3.2464 | 0.3080 | 2.5771 | 0.3880 | 3 |
| 4 | 1.3605 | 0.7350 | 4.5061 | 0.2219 | 3.3121 | 0.3019 | 4 |
| 5 | 1.4693 | 0.6806 | 5.8666 | 0.1705 | 3.9927 | 0.2505 | 5 |
| 6 | 1.5869 | 0.6302 | 7.3359 | 0.1363 | 4.6229 | 0.2163 | 6 |
| 7 | 1.7138 | 0.5835 | 8.9228 | 0.1121 | 5.2064 | 0.1921 | 7 |
| 8 | 1.8509 | 0.5403 | 10.6366 | 0.0940 | 5.7466 | 0.1740 | 8 |
| 9 | 1.9990 | 0.5002 | 12.4876 | 0.0801 | 6.2469 | 0.1601 | 9 |
| 10 | 2.1589 | 0.4632 | 14.4866 | 0.0690 | 6.7101 | 0.1490 | 10 |
| 11 | 2.3316 | 0.4289 | 16.6455 | 0.0601 | 7.1390 | 0.1401 | 11 |
| 12 | 2.5182 | 0.3971 | 18.9771 | 0.0527 | 7.5361 | 0.1327 | 12 |
| 13 | 2.7196 | 0.3677 | 21.4953 | 0.0465 | 7.9038 | 0.1265 | 13 |
| 14 | 2.9372 | 0.3405 | 24.2149 | 0.0413 | 8.2442 | 0.1213 | 14 |
| 15 | 3.1722 | 0.3152 | 27.1521 | 0.0368 | 8.5595 | 0.1168 | 15 |
| 16 | 3.4259 | 0.2919 | 30.3243 | 0.0330 | 8.8514 | 0.1130 | 16 |
| 17 | 3.7000 | 0.2703 | 33.7502 | 0.0296 | 9.1216 | 0.1096 | 17 |
| 18 | 3.9960 | 0.2502 | 37.4502 | 0.0267 | 9.3719 | 0.1067 | 18 |
| 19 | 4.3157 | 0.2317 | 41.4463 | 0.0241 | 9.6036 | 0.1041 | 19 |
| 20 | 4.6610 | 0.2145 | 45.7620 | 0.0219 | 9.8181 | 0.1019 | 20 |
| 21 | 5.0338 | 0.1987 | 50.4229 | 0.0198 | 10.0168 | 0.0998 | 21 |
| 22 | 5.4365 | 0.1839 | 55.4568 | 0.0180 | 10.2007 | 0.0980 | 22 |
| 23 | 5.8715 | 0.1703 | 60.8933 | 0.0164 | 10.3711 | 0.0964 | 23 |
| 24 | 6.3412 | 0.1577 | 66.7648 | 0.0150 | 10.5288 | 0.0950 | 24 |
| 25 | 6.8485 | 0.1460 | 73.1059 | 0.0137 | 10.6748 | 0.0937 | 25 |
| 26 | 7.3964 | 0.1352 | 79.9544 | 0.0125 | 10.8100 | 0.0925 | 26 |
| 27 | 7.9881 | 0.1252 | 87.3508 | 0.0114 | 10.9352 | 0.0914 | 27 |
| 28 | 8.6271 | 0.1159 | 95.3388 | 0.0105 | 11.0511 | 0.0905 | 28 |
| 29 | 9.3173 | 0.1073 | 103.9659 | 0.0096 | 11.1584 | 0.0896 | 29 |
| 30 | 10.0627 | 0.0994 | 113.2832 | 0.0088 | 11.2578 | 0.0888 | 30 |
| 31 | 10.8677 | 0.0920 | 123.3459 | 0.0081 | 11.3498 | 0.0881 | 31 |
| 32 | 11.7371 | 0.0852 | 134.2135 | 0.0075 | 11.4350 | 0.0875 | 32 |
| 33 | 12.6760 | 0.0789 | 145.9506 | 0.0069 | 11.5139 | 0.0869 | 33 |
| 34 | 13.6901 | 0.0730 | 158.6267 | 0.0063 | 11.5869 | 0.0863 | 34 |
| 35 | 14.7853 | 0.0676 | 172.3168 | 0.0058 | 11.6546 | 0.0858 | 35 |
| 36 | 15.9682 | 0.0626 | 187.1021 | 0.0053 | 11.7172 | 0.0853 | 36 |
| 37 | 17.2456 | 0.0580 | 203.0703 | 0.0049 | 11.7752 | 0.0849 | 37 |
| 38 | 18.6253 | 0.0537 | 220.3159 | 0.0045 | 11.8289 | 0.0845 | 38 |
| 39 | 20.1153 | 0.0497 | 238.9412 | 0.0042 | 11.8786 | 0.0842 | 39 |
| 40 | 21.7245 | 0.0460 | 259.0565 | 0.0039 | 11.9246 | 0.0839 | 40 |
| 50 | 46.9016 | 0.0213 | 573.7702 | 0.0017 | 12.2335 | 0.0817 | 50 |
| 60 | 101.2571 | 0.0099 | 1253.2133 | 0.0008 | 12.3766 | 0.0808 | 60 |
| 70 | 218.6064 | 0.0046 | 2720.0801 | 0.0004 | 12.4428 | 0.0804 | 70 |
| 80 | 471.9548 | 0.0021 | 5886.9354 | 0.0002 | 12.4735 | 0.0802 | 80 |
| 90 | 1018.9151 | 0.0010 | 12723.9386 | 0.0001 | 12.4877 | 0.0801 | 90 |
| 100 | 2199.7613 | 0.0005 | 27484.5157 | 0.0000 | 12.4943 | 0.0800 | 100 |
| 110 | 4749.1196 | 0.0002 | 59351.4946 | 0.0000 | 12.4974 | 0.0800 | 110 |
| 120 | 10252.9929 | 0.0001 | 128149.9118 | 0.0000 | 12.4988 | 0.0800 | 120 |

9.00% COMPOUND INTEREST FACTORS

| N | F/P | P/F | F/A | A/F | P/A | A/P | N |
|---|---|---|---|---|---|---|---|
| 1 | 1.0900 | 0.9174 | 1.0000 | 1.0000 | 0.9174 | 1.0900 | 1 |
| 2 | 1.1881 | 0.8417 | 2.0900 | 0.4785 | 1.7591 | 0.5685 | 2 |
| 3 | 1.2950 | 0.7722 | 3.2781 | 0.3051 | 2.5313 | 0.3951 | 3 |
| 4 | 1.4116 | 0.7084 | 4.5731 | 0.2187 | 3.2397 | 0.3087 | 4 |
| 5 | 1.5386 | 0.6499 | 5.9847 | 0.1671 | 3.8897 | 0.2571 | 5 |
| 6 | 1.6771 | 0.5963 | 7.5233 | 0.1329 | 4.4859 | 0.2229 | 6 |
| 7 | 1.8280 | 0.5470 | 9.2004 | 0.1087 | 5.0330 | 0.1987 | 7 |
| 8 | 1.9926 | 0.5019 | 11.0285 | 0.0907 | 5.5348 | 0.1807 | 8 |
| 9 | 2.1719 | 0.4604 | 13.0210 | 0.0768 | 5.9952 | 0.1668 | 9 |
| 10 | 2.3674 | 0.4224 | 15.1929 | 0.0658 | 6.4177 | 0.1558 | 10 |
| 11 | 2.5804 | 0.3875 | 17.5603 | 0.0569 | 6.8052 | 0.1469 | 11 |
| 12 | 2.8127 | 0.3555 | 20.1407 | 0.0497 | 7.1607 | 0.1397 | 12 |
| 13 | 3.0658 | 0.3262 | 22.9534 | 0.0436 | 7.4869 | 0.1336 | 13 |
| 14 | 3.3417 | 0.2992 | 26.0192 | 0.0384 | 7.7862 | 0.1284 | 14 |
| 15 | 3.6425 | 0.2745 | 29.3609 | 0.0341 | 8.0607 | 0.1241 | 15 |
| 16 | 3.9703 | 0.2519 | 33.0034 | 0.0303 | 8.3126 | 0.1203 | 16 |
| 17 | 4.3276 | 0.2311 | 36.9737 | 0.0270 | 8.5436 | 0.1170 | 17 |
| 18 | 4.7171 | 0.2120 | 41.3013 | 0.0242 | 8.7556 | 0.1142 | 18 |
| 19 | 5.1417 | 0.1945 | 46.0185 | 0.0217 | 8.9501 | 0.1117 | 19 |
| 20 | 5.6044 | 0.1784 | 51.1601 | 0.0195 | 9.1285 | 0.1095 | 20 |
| 21 | 6.1088 | 0.1637 | 56.7645 | 0.0176 | 9.2922 | 0.1076 | 21 |
| 22 | 6.6586 | 0.1502 | 62.8733 | 0.0159 | 9.4424 | 0.1059 | 22 |
| 23 | 7.2579 | 0.1378 | 69.5319 | 0.0144 | 9.5802 | 0.1044 | 23 |
| 24 | 7.9111 | 0.1264 | 76.7898 | 0.0130 | 9.7066 | 0.1030 | 24 |
| 25 | 8.6231 | 0.1160 | 84.7009 | 0.0118 | 9.8226 | 0.1018 | 25 |
| 26 | 9.3992 | 0.1064 | 93.3240 | 0.0107 | 9.9290 | 0.1007 | 26 |
| 27 | 10.2451 | 0.0976 | 102.7231 | 0.0097 | 10.0266 | 0.0997 | 27 |
| 28 | 11.1671 | 0.0895 | 112.9682 | 0.0089 | 10.1161 | 0.0989 | 28 |
| 29 | 12.1722 | 0.0822 | 124.1354 | 0.0081 | 10.1983 | 0.0981 | 29 |
| 30 | 13.2677 | 0.0754 | 136.3075 | 0.0073 | 10.2737 | 0.0973 | 30 |
| 31 | 14.4618 | 0.0691 | 149.5752 | 0.0067 | 10.3428 | 0.0967 | 31 |
| 32 | 15.7633 | 0.0634 | 164.0370 | 0.0061 | 10.4062 | 0.0961 | 32 |
| 33 | 17.1820 | 0.0582 | 179.8003 | 0.0056 | 10.4644 | 0.0956 | 33 |
| 34 | 18.7284 | 0.0534 | 196.9823 | 0.0051 | 10.5178 | 0.0951 | 34 |
| 35 | 20.4140 | 0.0490 | 215.7108 | 0.0046 | 10.5668 | 0.0946 | 35 |
| 36 | 22.2512 | 0.0449 | 236.1247 | 0.0042 | 10.6118 | 0.0942 | 36 |
| 37 | 24.2538 | 0.0412 | 258.3759 | 0.0039 | 10.6530 | 0.0939 | 37 |
| 38 | 26.4367 | 0.0378 | 282.6298 | 0.0035 | 10.6908 | 0.0935 | 38 |
| 39 | 28.8160 | 0.0347 | 309.0665 | 0.0032 | 10.7255 | 0.0932 | 39 |
| 40 | 31.4094 | 0.0318 | 337.8824 | 0.0030 | 10.7574 | 0.0930 | 40 |
| 50 | 74.3575 | 0.0134 | 815.0836 | 0.0012 | 10.9617 | 0.0912 | 50 |
| 60 | 176.0313 | 0.0057 | 1944.7921 | 0.0005 | 11.0480 | 0.0905 | 60 |
| 70 | 416.7301 | 0.0024 | 4619.2232 | 0.0002 | 11.0844 | 0.0902 | 70 |
| 80 | 986.5517 | 0.0010 | 10950.5741 | 0.0001 | 11.0998 | 0.0901 | 80 |
| 90 | 2335.5266 | 0.0004 | 25939.1842 | 0.0000 | 11.1064 | 0.0900 | 90 |
| 100 | 5529.0408 | 0.0002 | 61422.6755 | 0.0000 | 11.1091 | 0.0900 | 100 |
| 110 | 13089.2503 | 0.0001 | 145425.0036 | 0.0000 | 11.1103 | 0.0900 | 110 |
| 120 | 30987.0157 | 0.0000 | 344289.0639 | 0.0000 | 11.1108 | 0.0900 | 120 |

10.00% COMPOUND INTEREST FACTORS

| N | F/P | P/F | F/A | A/F | P/A | A/P | N |
|---|---|---|---|---|---|---|---|
| 1 | 1.1000 | 0.9091 | 1.0000 | 1.0000 | 0.9091 | 1.1000 | 1 |
| 2 | 1.2100 | 0.8264 | 2.1000 | 0.4762 | 1.7355 | 0.5762 | 2 |
| 3 | 1.3310 | 0.7513 | 3.3100 | 0.3021 | 2.4869 | 0.4021 | 3 |
| 4 | 1.4641 | 0.6830 | 4.6410 | 0.2155 | 3.1699 | 0.3155 | 4 |
| 5 | 1.6105 | 0.6209 | 6.1051 | 0.1638 | 3.7908 | 0.2638 | 5 |
| 6 | 1.7716 | 0.5645 | 7.7156 | 0.1296 | 4.3553 | 0.2296 | 6 |
| 7 | 1.9487 | 0.5132 | 9.4872 | 0.1054 | 4.8684 | 0.2054 | 7 |
| 8 | 2.1436 | 0.4665 | 11.4359 | 0.0874 | 5.3349 | 0.1874 | 8 |
| 9 | 2.3579 | 0.4241 | 13.5795 | 0.0736 | 5.7590 | 0.1736 | 9 |
| 10 | 2.5937 | 0.3855 | 15.9374 | 0.0627 | 6.1446 | 0.1627 | 10 |
| 11 | 2.8531 | 0.3505 | 18.5312 | 0.0540 | 6.4951 | 0.1540 | 11 |
| 12 | 3.1384 | 0.3186 | 21.3843 | 0.0468 | 6.8137 | 0.1468 | 12 |
| 13 | 3.4523 | 0.2897 | 24.5227 | 0.0408 | 7.1034 | 0.1408 | 13 |
| 14 | 3.7975 | 0.2633 | 27.9750 | 0.0357 | 7.3667 | 0.1357 | 14 |
| 15 | 4.1772 | 0.2394 | 31.7725 | 0.0315 | 7.6061 | 0.1315 | 15 |
| 16 | 4.5950 | 0.2176 | 35.9497 | 0.0278 | 7.8237 | 0.1278 | 16 |
| 17 | 5.0545 | 0.1978 | 40.5447 | 0.0247 | 8.0216 | 0.1247 | 17 |
| 18 | 5.5599 | 0.1799 | 45.5992 | 0.0219 | 8.2014 | 0.1219 | 18 |
| 19 | 6.1159 | 0.1635 | 51.1591 | 0.0195 | 8.3649 | 0.1195 | 19 |
| 20 | 6.7275 | 0.1486 | 57.2750 | 0.0175 | 8.5136 | 0.1175 | 20 |
| 21 | 7.4002 | 0.1351 | 64.0025 | 0.0156 | 8.6487 | 0.1156 | 21 |
| 22 | 8.1403 | 0.1228 | 71.4027 | 0.0140 | 8.7715 | 0.1140 | 22 |
| 23 | 8.9543 | 0.1117 | 79.5430 | 0.0126 | 8.8832 | 0.1126 | 23 |
| 24 | 9.8497 | 0.1015 | 88.4973 | 0.0113 | 8.9847 | 0.1113 | 24 |
| 25 | 10.8347 | 0.0923 | 98.3471 | 0.0102 | 9.0770 | 0.1102 | 25 |
| 26 | 11.9182 | 0.0839 | 109.1818 | 0.0092 | 9.1609 | 0.1092 | 26 |
| 27 | 13.1100 | 0.0763 | 121.0999 | 0.0083 | 9.2372 | 0.1083 | 27 |
| 28 | 14.4210 | 0.0693 | 134.2099 | 0.0075 | 9.3066 | 0.1075 | 28 |
| 29 | 15.8631 | 0.0630 | 148.6309 | 0.0067 | 9.3696 | 0.1067 | 29 |
| 30 | 17.4494 | 0.0573 | 164.4940 | 0.0061 | 9.4269 | 0.1061 | 30 |
| 31 | 19.1943 | 0.0521 | 181.9434 | 0.0055 | 9.4790 | 0.1055 | 31 |
| 32 | 21.1138 | 0.0474 | 201.1378 | 0.0050 | 9.5264 | 0.1050 | 32 |
| 33 | 23.2252 | 0.0431 | 222.2515 | 0.0045 | 9.5694 | 0.1045 | 33 |
| 34 | 25.5477 | 0.0391 | 245.4767 | 0.0041 | 9.6086 | 0.1041 | 34 |
| 35 | 28.1024 | 0.0356 | 271.0244 | 0.0037 | 9.6442 | 0.1037 | 35 |
| 36 | 30.9127 | 0.0323 | 299.1268 | 0.0033 | 9.6765 | 0.1033 | 36 |
| 37 | 34.0039 | 0.0294 | 330.0395 | 0.0030 | 9.7059 | 0.1030 | 37 |
| 38 | 37.4043 | 0.0267 | 364.0434 | 0.0027 | 9.7327 | 0.1027 | 38 |
| 39 | 41.1448 | 0.0243 | 401.4478 | 0.0025 | 9.7570 | 0.1025 | 39 |
| 40 | 45.2593 | 0.0221 | 442.5926 | 0.0023 | 9.7791 | 0.1023 | 40 |
| 50 | 117.3909 | 0.0085 | 1163.9085 | 0.0009 | 9.9148 | 0.1009 | 50 |
| 60 | 304.4816 | 0.0033 | 3034.8164 | 0.0003 | 9.9672 | 0.1003 | 60 |
| 70 | 789.7470 | 0.0013 | 7887.4696 | 0.0001 | 9.9873 | 0.1001 | 70 |
| 80 | 2048.4002 | 0.0005 | 20474.0021 | 0.0000 | 9.9951 | 0.1000 | 80 |
| 90 | 5313.0226 | 0.0002 | 53120.2261 | 0.0000 | 9.9981 | 0.1000 | 90 |
| 100 | 13780.6123 | 0.0001 | 137796.1234 | 0.0000 | 9.9993 | 0.1000 | 100 |
| 110 | 35743.3594 | 0.0000 | 357423.5935 | 0.0000 | 9.9997 | 0.1000 | 110 |
| 120 | 92709.0688 | 0.0000 | 927080.6882 | 0.0000 | 9.9999 | 0.1000 | 120 |

12.00% COMPOUND INTEREST FACTORS

| N | F/P | P/F | F/A | A/F | P/A | A/P | N |
|---|---|---|---|---|---|---|---|
| 1 | 1.1200 | 0.8929 | 1.0000 | 1.0000 | 0.8929 | 1.1200 | 1 |
| 2 | 1.2544 | 0.7972 | 2.1200 | 0.4717 | 1.6901 | 0.5917 | 2 |
| 3 | 1.4049 | 0.7118 | 3.3744 | 0.2963 | 2.4018 | 0.4163 | 3 |
| 4 | 1.5735 | 0.6355 | 4.7793 | 0.2092 | 3.0373 | 0.3292 | 4 |
| 5 | 1.7623 | 0.5674 | 6.3528 | 0.1574 | 3.6048 | 0.2774 | 5 |
| 6 | 1.9738 | 0.5066 | 8.1152 | 0.1232 | 4.1114 | 0.2432 | 6 |
| 7 | 2.2107 | 0.4523 | 10.0890 | 0.0991 | 4.5638 | 0.2191 | 7 |
| 8 | 2.4760 | 0.4039 | 12.2997 | 0.0813 | 4.9676 | 0.2013 | 8 |
| 9 | 2.7731 | 0.3606 | 14.7757 | 0.0677 | 5.3282 | 0.1877 | 9 |
| 10 | 3.1058 | 0.3220 | 17.5487 | 0.0570 | 5.6502 | 0.1770 | 10 |
| 11 | 3.4785 | 0.2875 | 20.6546 | 0.0484 | 5.9377 | 0.1684 | 11 |
| 12 | 3.8960 | 0.2567 | 24.1331 | 0.0414 | 6.1944 | 0.1614 | 12 |
| 13 | 4.3635 | 0.2292 | 28.0291 | 0.0357 | 6.4235 | 0.1557 | 13 |
| 14 | 4.8871 | 0.2046 | 32.3926 | 0.0309 | 6.6282 | 0.1509 | 14 |
| 15 | 5.4736 | 0.1827 | 37.2797 | 0.0268 | 6.8109 | 0.1468 | 15 |
| 16 | 6.1304 | 0.1631 | 42.7533 | 0.0234 | 6.9740 | 0.1434 | 16 |
| 17 | 6.8660 | 0.1456 | 48.8837 | 0.0205 | 7.1196 | 0.1405 | 17 |
| 18 | 7.6900 | 0.1300 | 55.7497 | 0.0179 | 7.2497 | 0.1379 | 18 |
| 19 | 8.6128 | 0.1161 | 63.4397 | 0.0158 | 7.3658 | 0.1358 | 19 |
| 20 | 9.6463 | 0.1037 | 72.0524 | 0.0139 | 7.4694 | 0.1339 | 20 |
| 21 | 10.8038 | 0.0926 | 81.6987 | 0.0122 | 7.5620 | 0.1322 | 21 |
| 22 | 12.1003 | 0.0826 | 92.5026 | 0.0108 | 7.6446 | 0.1308 | 22 |
| 23 | 13.5523 | 0.0738 | 104.6029 | 0.0096 | 7.7184 | 0.1296 | 23 |
| 24 | 15.1786 | 0.0659 | 118.1552 | 0.0085 | 7.7843 | 0.1285 | 24 |
| 25 | 17.0001 | 0.0588 | 133.3339 | 0.0075 | 7.8431 | 0.1275 | 25 |
| 26 | 19.0401 | 0.0525 | 150.3339 | 0.0067 | 7.8957 | 0.1267 | 26 |
| 27 | 21.3249 | 0.0469 | 169.3740 | 0.0059 | 7.9426 | 0.1259 | 27 |
| 28 | 23.8839 | 0.0419 | 190.6989 | 0.0052 | 7.9844 | 0.1252 | 28 |
| 29 | 26.7499 | 0.0374 | 214.5828 | 0.0047 | 8.0218 | 0.1247 | 29 |
| 30 | 29.9599 | 0.0334 | 241.3327 | 0.0041 | 8.0552 | 0.1241 | 30 |
| 31 | 33.5551 | 0.0298 | 271.2926 | 0.0037 | 8.0850 | 0.1237 | 31 |
| 32 | 37.5817 | 0.0266 | 304.8477 | 0.0033 | 8.1116 | 0.1233 | 32 |
| 33 | 42.0915 | 0.0238 | 342.4294 | 0.0029 | 8.1354 | 0.1229 | 33 |
| 34 | 47.1425 | 0.0212 | 384.5210 | 0.0026 | 8.1566 | 0.1226 | 34 |
| 35 | 52.7996 | 0.0189 | 431.6635 | 0.0023 | 8.1755 | 0.1223 | 35 |
| 36 | 59.1356 | 0.0169 | 484.4631 | 0.0021 | 8.1924 | 0.1221 | 36 |
| 37 | 66.2318 | 0.0151 | 543.5987 | 0.0018 | 8.2075 | 0.1218 | 37 |
| 38 | 74.1797 | 0.0135 | 609.8305 | 0.0016 | 8.2210 | 0.1216 | 38 |
| 39 | 83.0812 | 0.0120 | 684.0102 | 0.0015 | 8.2330 | 0.1215 | 39 |
| 40 | 93.0510 | 0.0107 | 767.0914 | 0.0013 | 8.2438 | 0.1213 | 40 |
| 50 | 289.0022 | 0.0035 | 2400.0182 | 0.0004 | 8.3045 | 0.1204 | 50 |
| 60 | 897.5969 | 0.0011 | 7471.6411 | 0.0001 | 8.3240 | 0.1201 | 60 |
| 70 | 2787.7998 | 0.0004 | 23223.3319 | 0.0000 | 8.3303 | 0.1200 | 70 |
| 80 | 8658.4831 | 0.0001 | 72145.6925 | 0.0000 | 8.3324 | 0.1200 | 80 |
| 90 | 26891.9342 | 0.0000 | 224091.1185 | 0.0000 | 8.3330 | 0.1200 | 90 |
| 100 | 83522.2657 | 0.0000 | 696010.5477 | 0.0000 | 8.3332 | 0.1200 | 100 |

15.00% COMPOUND INTEREST FACTORS

| N | F/P | P/F | F/A | A/F | P/A | A/P | N |
|---|---|---|---|---|---|---|---|
| 1 | 1.1500 | 0.8696 | 1.0000 | 1.0000 | 0.8696 | 1.1500 | 1 |
| 2 | 1.3225 | 0.7561 | 2.1500 | 0.4651 | 1.6257 | 0.6151 | 2 |
| 3 | 1.5209 | 0.6575 | 3.4725 | 0.2880 | 2.2832 | 0.4380 | 3 |
| 4 | 1.7490 | 0.5718 | 4.9934 | 0.2003 | 2.8550 | 0.3503 | 4 |
| 5 | 2.0114 | 0.4972 | 6.7424 | 0.1483 | 3.3522 | 0.2983 | 5 |
| 6 | 2.3131 | 0.4323 | 8.7537 | 0.1142 | 3.7845 | 0.2642 | 6 |
| 7 | 2.6600 | 0.3759 | 11.0668 | 0.0904 | 4.1604 | 0.2404 | 7 |
| 8 | 3.0590 | 0.3269 | 13.7268 | 0.0729 | 4.4873 | 0.2229 | 8 |
| 9 | 3.5179 | 0.2843 | 16.7858 | 0.0596 | 4.7716 | 0.2096 | 9 |
| 10 | 4.0456 | 0.2472 | 20.3037 | 0.0493 | 5.0188 | 0.1993 | 10 |
| 11 | 4.6524 | 0.2149 | 24.3493 | 0.0411 | 5.2337 | 0.1911 | 11 |
| 12 | 5.3503 | 0.1869 | 29.0017 | 0.0345 | 5.4206 | 0.1845 | 12 |
| 13 | 6.1528 | 0.1625 | 34.3519 | 0.0291 | 5.5831 | 0.1791 | 13 |
| 14 | 7.0757 | 0.1413 | 40.5047 | 0.0247 | 5.7245 | 0.1747 | 14 |
| 15 | 8.1371 | 0.1229 | 47.5804 | 0.0210 | 5.8474 | 0.1710 | 15 |
| 16 | 9.3576 | 0.1069 | 55.7175 | 0.0179 | 5.9542 | 0.1679 | 16 |
| 17 | 10.7613 | 0.0929 | 65.0751 | 0.0154 | 6.0472 | 0.1654 | 17 |
| 18 | 12.3755 | 0.0808 | 75.8364 | 0.0132 | 6.1280 | 0.1632 | 18 |
| 19 | 14.2318 | 0.0703 | 88.2118 | 0.0113 | 6.1982 | 0.1613 | 19 |
| 20 | 16.3665 | 0.0611 | 102.4436 | 0.0098 | 6.2593 | 0.1598 | 20 |
| 21 | 18.8215 | 0.0531 | 118.8101 | 0.0084 | 6.3125 | 0.1584 | 21 |
| 22 | 21.6447 | 0.0462 | 137.6316 | 0.0073 | 6.3587 | 0.1573 | 22 |
| 23 | 24.8915 | 0.0402 | 159.2764 | 0.0063 | 6.3988 | 0.1563 | 23 |
| 24 | 28.6252 | 0.0349 | 184.1678 | 0.0054 | 6.4338 | 0.1554 | 24 |
| 25 | 32.9190 | 0.0304 | 212.7930 | 0.0047 | 6.4641 | 0.1547 | 25 |
| 26 | 37.8568 | 0.0264 | 245.7120 | 0.0041 | 6.4906 | 0.1541 | 26 |
| 27 | 43.5353 | 0.0230 | 283.5688 | 0.0035 | 6.5135 | 0.1535 | 27 |
| 28 | 50.0656 | 0.0200 | 327.1041 | 0.0031 | 6.5335 | 0.1531 | 28 |
| 29 | 57.5755 | 0.0174 | 377.1697 | 0.0027 | 6.5509 | 0.1527 | 29 |
| 30 | 66.2118 | 0.0151 | 434.7451 | 0.0023 | 6.5660 | 0.1523 | 30 |
| 31 | 76.1435 | 0.0131 | 500.9569 | 0.0020 | 6.5791 | 0.1520 | 31 |
| 32 | 87.5651 | 0.0114 | 577.1005 | 0.0017 | 6.5905 | 0.1517 | 32 |
| 33 | 100.6998 | 0.0099 | 664.6655 | 0.0015 | 6.6005 | 0.1515 | 33 |
| 34 | 115.8048 | 0.0086 | 765.3654 | 0.0013 | 6.6091 | 0.1513 | 34 |
| 35 | 133.1755 | 0.0075 | 881.1702 | 0.0011 | 6.6166 | 0.1511 | 35 |
| 36 | 153.1519 | 0.0065 | 1014.3457 | 0.0010 | 6.6231 | 0.1510 | 36 |
| 37 | 176.1246 | 0.0057 | 1167.4975 | 0.0009 | 6.6288 | 0.1509 | 37 |
| 38 | 202.5433 | 0.0049 | 1343.6222 | 0.0007 | 6.6338 | 0.1507 | 38 |
| 39 | 232.9248 | 0.0043 | 1546.1655 | 0.0006 | 6.6380 | 0.1506 | 39 |
| 40 | 267.8635 | 0.0037 | 1779.0903 | 0.0006 | 6.6418 | 0.1506 | 40 |
| 50 | 1083.6574 | 0.0009 | 7217.7163 | 0.0001 | 6.6605 | 0.1501 | 50 |
| 60 | 4383.9987 | 0.0002 | 29219.9916 | 0.0000 | 6.6651 | 0.1500 | 60 |
| 70 | 17735.7200 | 0.0001 | 118231.4669 | 0.0000 | 6.6663 | 0.1500 | 70 |
| 80 | 71750.8794 | 0.0000 | 478332.5293 | 0.0000 | 6.6666 | 0.1500 | 80 |

20.00% COMPOUND INTEREST FACTORS

| N | F/P | P/F | F/A | A/F | P/A | A/P | N |
|---|---|---|---|---|---|---|---|
| 1 | 1.2000 | 0.8333 | 1.0000 | 1.0000 | 0.8333 | 1.2000 | 1 |
| 2 | 1.4400 | 0.6944 | 2.2000 | 0.4545 | 1.5278 | 0.6545 | 2 |
| 3 | 1.7280 | 0.5787 | 3.6400 | 0.2747 | 2.1065 | 0.4747 | 3 |
| 4 | 2.0736 | 0.4823 | 5.3680 | 0.1863 | 2.5887 | 0.3863 | 4 |
| 5 | 2.4883 | 0.4019 | 7.4416 | 0.1344 | 2.9906 | 0.3344 | 5 |
| 6 | 2.9860 | 0.3349 | 9.9299 | 0.1007 | 3.3255 | 0.3007 | 6 |
| 7 | 3.5832 | 0.2791 | 12.9159 | 0.0774 | 3.6046 | 0.2774 | 7 |
| 8 | 4.2998 | 0.2326 | 16.4991 | 0.0606 | 3.8372 | 0.2606 | 8 |
| 9 | 5.1598 | 0.1938 | 20.7989 | 0.0481 | 4.0310 | 0.2481 | 9 |
| 10 | 6.1917 | 0.1615 | 25.9587 | 0.0385 | 4.1925 | 0.2385 | 10 |
| 11 | 7.4301 | 0.1346 | 32.1504 | 0.0311 | 4.3271 | 0.2311 | 11 |
| 12 | 8.9161 | 0.1122 | 39.5805 | 0.0253 | 4.4392 | 0.2253 | 12 |
| 13 | 10.6993 | 0.0935 | 48.4966 | 0.0206 | 4.5327 | 0.2206 | 13 |
| 14 | 12.8392 | 0.0779 | 59.1959 | 0.0169 | 4.6106 | 0.2169 | 14 |
| 15 | 15.4070 | 0.0649 | 72.0351 | 0.0139 | 4.6755 | 0.2139 | 15 |
| 16 | 18.4884 | 0.0541 | 87.4421 | 0.0114 | 4.7296 | 0.2114 | 16 |
| 17 | 22.1861 | 0.0451 | 105.9306 | 0.0094 | 4.7746 | 0.2094 | 17 |
| 18 | 26.6233 | 0.0376 | 128.1167 | 0.0078 | 4.8122 | 0.2078 | 18 |
| 19 | 31.9480 | 0.0313 | 154.7400 | 0.0065 | 4.8435 | 0.2065 | 19 |
| 20 | 38.3376 | 0.0261 | 186.6880 | 0.0054 | 4.8696 | 0.2054 | 20 |
| 21 | 46.0051 | 0.0217 | 225.0256 | 0.0044 | 4.8913 | 0.2044 | 21 |
| 22 | 55.2061 | 0.0181 | 271.0307 | 0.0037 | 4.9094 | 0.2037 | 22 |
| 23 | 66.2474 | 0.0151 | 326.2369 | 0.0031 | 4.9245 | 0.2031 | 23 |
| 24 | 79.4968 | 0.0126 | 392.4842 | 0.0025 | 4.9371 | 0.2025 | 24 |
| 25 | 95.3962 | 0.0105 | 471.9811 | 0.0021 | 4.9476 | 0.2021 | 25 |
| 26 | 114.4755 | 0.0087 | 567.3773 | 0.0018 | 4.9563 | 0.2018 | 26 |
| 27 | 137.3706 | 0.0073 | 681.8528 | 0.0015 | 4.9636 | 0.2015 | 27 |
| 28 | 164.8447 | 0.0061 | 819.2233 | 0.0012 | 4.9697 | 0.2012 | 28 |
| 29 | 197.8136 | 0.0051 | 984.0680 | 0.0010 | 4.9747 | 0.2010 | 29 |
| 30 | 237.3763 | 0.0042 | 1181.8816 | 0.0008 | 4.9789 | 0.2008 | 30 |
| 31 | 284.8516 | 0.0035 | 1419.2579 | 0.0007 | 4.9824 | 0.2007 | 31 |
| 32 | 341.8219 | 0.0029 | 1704.1095 | 0.0006 | 4.9854 | 0.2006 | 32 |
| 33 | 410.1863 | 0.0024 | 2045.9314 | 0.0005 | 4.9878 | 0.2005 | 33 |
| 34 | 492.2235 | 0.0020 | 2456.1176 | 0.0004 | 4.9898 | 0.2004 | 34 |
| 35 | 590.6682 | 0.0017 | 2948.3411 | 0.0003 | 4.9915 | 0.2003 | 35 |
| 36 | 708.8019 | 0.0014 | 3539.0094 | 0.0003 | 4.9929 | 0.2003 | 36 |
| 37 | 850.5622 | 0.0012 | 4247.8112 | 0.0002 | 4.9941 | 0.2002 | 37 |
| 38 | 1020.6747 | 0.0010 | 5098.3735 | 0.0002 | 4.9951 | 0.2002 | 38 |
| 39 | 1224.8096 | 0.0008 | 6119.0482 | 0.0002 | 4.9959 | 0.2002 | 39 |
| 40 | 1469.7716 | 0.0007 | 7343.8578 | 0.0001 | 4.9966 | 0.2001 | 40 |
| 50 | 9100.4382 | 0.0001 | 45497.1908 | 0.0000 | 4.9995 | 0.2000 | 50 |

25.00% COMPOUND INTEREST FACTORS

| N | F/P | P/F | F/A | A/F | P/A | A/P | N |
|---|---|---|---|---|---|---|---|
| 1 | 1.2500 | 0.8000 | 1.0000 | 1.0000 | 0.8000 | 1.2500 | 1 |
| 2 | 1.5625 | 0.6400 | 2.2500 | 0.4444 | 1.4400 | 0.6944 | 2 |
| 3 | 1.9531 | 0.5120 | 3.8125 | 0.2623 | 1.9520 | 0.5123 | 3 |
| 4 | 2.4414 | 0.4096 | 5.7656 | 0.1734 | 2.3616 | 0.4234 | 4 |
| 5 | 3.0518 | 0.3277 | 8.2070 | 0.1218 | 2.6893 | 0.3718 | 5 |
| 6 | 3.8147 | 0.2621 | 11.2588 | 0.0888 | 2.9514 | 0.3388 | 6 |
| 7 | 4.7684 | 0.2097 | 15.0735 | 0.0663 | 3.1611 | 0.3163 | 7 |
| 8 | 5.9605 | 0.1678 | 19.8419 | 0.0504 | 3.3289 | 0.3004 | 8 |
| 9 | 7.4506 | 0.1342 | 25.8023 | 0.0388 | 3.4631 | 0.2888 | 9 |
| 10 | 9.3132 | 0.1074 | 33.2529 | 0.0301 | 3.5705 | 0.2801 | 10 |
| 11 | 11.6415 | 0.0859 | 42.5661 | 0.0235 | 3.6564 | 0.2735 | 11 |
| 12 | 14.5519 | 0.0687 | 54.2077 | 0.0184 | 3.7251 | 0.2684 | 12 |
| 13 | 18.1899 | 0.0550 | 68.7596 | 0.0145 | 3.7801 | 0.2645 | 13 |
| 14 | 22.7374 | 0.0440 | 86.9495 | 0.0115 | 3.8241 | 0.2615 | 14 |
| 15 | 28.4217 | 0.0352 | 109.6868 | 0.0091 | 3.8593 | 0.2591 | 15 |
| 16 | 35.5271 | 0.0281 | 138.1085 | 0.0072 | 3.8874 | 0.2572 | 16 |
| 17 | 44.4089 | 0.0225 | 173.6357 | 0.0058 | 3.9099 | 0.2558 | 17 |
| 18 | 55.5112 | 0.0180 | 218.0446 | 0.0046 | 3.9279 | 0.2546 | 18 |
| 19 | 69.3889 | 0.0144 | 273.5558 | 0.0037 | 3.9424 | 0.2537 | 19 |
| 20 | 86.7362 | 0.0115 | 342.9447 | 0.0029 | 3.9539 | 0.2529 | 20 |
| 21 | 108.4202 | 0.0092 | 429.6809 | 0.0023 | 3.9631 | 0.2523 | 21 |
| 22 | 135.5253 | 0.0074 | 538.1011 | 0.0019 | 3.9705 | 0.2519 | 22 |
| 23 | 169.4066 | 0.0059 | 673.6264 | 0.0015 | 3.9764 | 0.2515 | 23 |
| 24 | 211.7582 | 0.0047 | 843.0329 | 0.0012 | 3.9811 | 0.2512 | 24 |
| 25 | 264.6978 | 0.0038 | 1054.7912 | 0.0009 | 3.9849 | 0.2509 | 25 |
| 26 | 330.8722 | 0.0030 | 1319.4890 | 0.0008 | 3.9879 | 0.2508 | 26 |
| 27 | 413.5903 | 0.0024 | 1650.3612 | 0.0006 | 3.9903 | 0.2506 | 27 |
| 28 | 516.9879 | 0.0019 | 2063.9515 | 0.0005 | 3.9923 | 0.2505 | 28 |
| 29 | 646.2349 | 0.0015 | 2580.9394 | 0.0004 | 3.9938 | 0.2504 | 29 |
| 30 | 807.7936 | 0.0012 | 3227.1743 | 0.0003 | 3.9950 | 0.2503 | 30 |
| 31 | 1009.7420 | 0.0010 | 4034.9678 | 0.0002 | 3.9960 | 0.2502 | 31 |
| 32 | 1262.1774 | 0.0008 | 5044.7098 | 0.0002 | 3.9968 | 0.2502 | 32 |
| 33 | 1577.7218 | 0.0006 | 6306.8872 | 0.0002 | 3.9975 | 0.2502 | 33 |
| 34 | 1972.1523 | 0.0005 | 7884.6091 | 0.0001 | 3.9980 | 0.2501 | 34 |
| 35 | 2465.1903 | 0.0004 | 9856.7613 | 0.0001 | 3.9984 | 0.2501 | 35 |
| 36 | 3081.4879 | 0.0003 | 12321.9516 | 0.0001 | 3.9987 | 0.2501 | 36 |
| 37 | 3851.8599 | 0.0003 | 15403.4396 | 0.0001 | 3.9990 | 0.2501 | 37 |
| 38 | 4814.8249 | 0.0002 | 19255.2994 | 0.0001 | 3.9992 | 0.2501 | 38 |
| 39 | 6018.5311 | 0.0002 | 24070.1243 | 0.0000 | 3.9993 | 0.2500 | 39 |
| 40 | 7523.1638 | 0.0001 | 30088.6554 | 0.0000 | 3.9995 | 0.2500 | 40 |
| 50 | 70064.9232 | 0.0000 | 280255.6929 | 0.0000 | 3.9999 | 0.2500 | 50 |

30.00% COMPOUND INTEREST FACTORS

| N | F/P | P/F | F/A | A/F | P/A | A/P | N |
|---|---|---|---|---|---|---|---|
| 1 | 1.3000 | 0.7692 | 1.0000 | 1.0000 | 0.7692 | 1.3000 | 1 |
| 2 | 1.6900 | 0.5917 | 2.3000 | 0.4348 | 1.3609 | 0.7348 | 2 |
| 3 | 2.1970 | 0.4552 | 3.9900 | 0.2506 | 1.8161 | 0.5506 | 3 |
| 4 | 2.8561 | 0.3501 | 6.1870 | 0.1616 | 2.1662 | 0.4616 | 4 |
| 5 | 3.7129 | 0.2693 | 9.0431 | 0.1106 | 2.4356 | 0.4106 | 5 |
| 6 | 4.8268 | 0.2072 | 12.7560 | 0.0784 | 2.6427 | 0.3784 | 6 |
| 7 | 6.2749 | 0.1594 | 17.5828 | 0.0569 | 2.8021 | 0.3569 | 7 |
| 8 | 8.1573 | 0.1226 | 23.8577 | 0.0419 | 2.9247 | 0.3419 | 8 |
| 9 | 10.6045 | 0.0943 | 32.0150 | 0.0312 | 3.0190 | 0.3312 | 9 |
| 10 | 13.7858 | 0.0725 | 42.6195 | 0.0235 | 3.0915 | 0.3235 | 10 |
| 11 | 17.9216 | 0.0558 | 56.4053 | 0.0177 | 3.1473 | 0.3177 | 11 |
| 12 | 23.2981 | 0.0429 | 74.3270 | 0.0135 | 3.1903 | 0.3135 | 12 |
| 13 | 30.2875 | 0.0330 | 97.6250 | 0.0102 | 3.2233 | 0.3102 | 13 |
| 14 | 39.3738 | 0.0254 | 127.9125 | 0.0078 | 3.2487 | 0.3078 | 14 |
| 15 | 51.1859 | 0.0195 | 167.2863 | 0.0060 | 3.2682 | 0.3060 | 15 |
| 16 | 66.5417 | 0.0150 | 218.4722 | 0.0046 | 3.2832 | 0.3046 | 16 |
| 17 | 86.5042 | 0.0116 | 285.0139 | 0.0035 | 3.2948 | 0.3035 | 17 |
| 18 | 112.4554 | 0.0089 | 371.5180 | 0.0027 | 3.3037 | 0.3027 | 18 |
| 19 | 146.1920 | 0.0068 | 483.9734 | 0.0021 | 3.3105 | 0.3021 | 19 |
| 20 | 190.0496 | 0.0053 | 630.1655 | 0.0016 | 3.3158 | 0.3016 | 20 |
| 21 | 247.0645 | 0.0040 | 820.2151 | 0.0012 | 3.3198 | 0.3012 | 21 |
| 22 | 321.1839 | 0.0031 | 1067.2796 | 0.0009 | 3.3230 | 0.3009 | 22 |
| 23 | 417.5391 | 0.0024 | 1388.4635 | 0.0007 | 3.3254 | 0.3007 | 23 |
| 24 | 542.8008 | 0.0018 | 1806.0026 | 0.0006 | 3.3272 | 0.3006 | 24 |
| 25 | 705.6410 | 0.0014 | 2348.8033 | 0.0004 | 3.3286 | 0.3004 | 25 |
| 26 | 917.3333 | 0.0011 | 3054.4443 | 0.0003 | 3.3297 | 0.3003 | 26 |
| 27 | 1192.5333 | 0.0008 | 3971.7776 | 0.0003 | 3.3305 | 0.3003 | 27 |
| 28 | 1550.2933 | 0.0006 | 5164.3109 | 0.0002 | 3.3312 | 0.3002 | 28 |
| 29 | 2015.3813 | 0.0005 | 6714.6042 | 0.0001 | 3.3317 | 0.3001 | 29 |
| 30 | 2619.9956 | 0.0004 | 8729.9855 | 0.0001 | 3.3321 | 0.3001 | 30 |
| 31 | 3405.9943 | 0.0003 | 11349.9811 | 0.0001 | 3.3324 | 0.3001 | 31 |
| 32 | 4427.7926 | 0.0002 | 14755.9755 | 0.0001 | 3.3326 | 0.3001 | 32 |
| 33 | 5756.1304 | 0.0002 | 19183.7681 | 0.0001 | 3.3328 | 0.3001 | 33 |
| 34 | 7482.9696 | 0.0001 | 24939.8985 | 0.0000 | 3.3329 | 0.3000 | 34 |
| 35 | 9727.8604 | 0.0001 | 32422.8681 | 0.0000 | 3.3330 | 0.3000 | 35 |
| 36 | 12646.2186 | 0.0001 | 42150.7285 | 0.0000 | 3.3331 | 0.3000 | 36 |
| 37 | 16440.0841 | 0.0001 | 54796.9471 | 0.0000 | 3.3331 | 0.3000 | 37 |
| 38 | 21372.1094 | 0.0000 | 71237.0312 | 0.0000 | 3.3332 | 0.3000 | 38 |
| 39 | 27783.7422 | 0.0000 | 92609.1405 | 0.0000 | 3.3332 | 0.3000 | 39 |
| 40 | 36118.8648 | 0.0000 | 120392.8827 | 0.0000 | 3.3332 | 0.3000 | 40 |

40.00% COMPOUND INTEREST FACTORS PAGE 26

| N | F/P | P/F | F/A | A/F | P/A | A/P | N |
|---|---|---|---|---|---|---|---|
| 1 | 1.4000 | 0.7143 | 1.0000 | 1.0000 | 0.7143 | 1.4000 | 1 |
| 2 | 1.9600 | 0.5102 | 2.4000 | 0.4167 | 1.2245 | 0.8167 | 2 |
| 3 | 2.7440 | 0.3644 | 4.3600 | 0.2294 | 1.5889 | 0.6294 | 3 |
| 4 | 3.8416 | 0.2603 | 7.1040 | 0.1408 | 1.8492 | 0.5408 | 4 |
| 5 | 5.3782 | 0.1859 | 10.9456 | 0.0914 | 2.0352 | 0.4914 | 5 |
| 6 | 7.5295 | 0.1328 | 16.3238 | 0.0613 | 2.1680 | 0.4613 | 6 |
| 7 | 10.5414 | 0.0949 | 23.8534 | 0.0419 | 2.2628 | 0.4419 | 7 |
| 8 | 14.7579 | 0.0678 | 34.3947 | 0.0291 | 2.3306 | 0.4291 | 8 |
| 9 | 20.6610 | 0.0484 | 49.1526 | 0.0203 | 2.3790 | 0.4203 | 9 |
| 10 | 28.9255 | 0.0346 | 69.8137 | 0.0143 | 2.4136 | 0.4143 | 10 |
| 11 | 40.4957 | 0.0247 | 98.7391 | 0.0101 | 2.4383 | 0.4101 | 11 |
| 12 | 56.6939 | 0.0176 | 139.2348 | 0.0072 | 2.4559 | 0.4072 | 12 |
| 13 | 79.3715 | 0.0126 | 195.9287 | 0.0051 | 2.4685 | 0.4051 | 13 |
| 14 | 111.1201 | 0.0090 | 275.3002 | 0.0036 | 2.4775 | 0.4036 | 14 |
| 15 | 155.5681 | 0.0064 | 386.4202 | 0.0026 | 2.4839 | 0.4026 | 15 |
| 16 | 217.7953 | 0.0046 | 541.9883 | 0.0018 | 2.4885 | 0.4018 | 16 |
| 17 | 304.9135 | 0.0033 | 759.7837 | 0.0013 | 2.4918 | 0.4013 | 17 |
| 18 | 426.8789 | 0.0023 | 1064.6971 | 0.0009 | 2.4941 | 0.4009 | 18 |
| 19 | 597.6304 | 0.0017 | 1491.5760 | 0.0007 | 2.4958 | 0.4007 | 19 |
| 20 | 836.6826 | 0.0012 | 2089.2064 | 0.0005 | 2.4970 | 0.4005 | 20 |
| 21 | 1171.3556 | 0.0009 | 2925.8889 | 0.0003 | 2.4979 | 0.4003 | 21 |
| 22 | 1639.8978 | 0.0006 | 4097.2445 | 0.0002 | 2.4985 | 0.4002 | 22 |
| 23 | 2295.8569 | 0.0004 | 5737.1423 | 0.0002 | 2.4989 | 0.4002 | 23 |
| 24 | 3214.1997 | 0.0003 | 8032.9993 | 0.0001 | 2.4992 | 0.4001 | 24 |
| 25 | 4499.8796 | 0.0002 | 11247.1990 | 0.0001 | 2.4994 | 0.4001 | 25 |
| 26 | 6299.8314 | 0.0002 | 15747.0785 | 0.0001 | 2.4996 | 0.4001 | 26 |
| 27 | 8819.7640 | 0.0001 | 22046.9099 | 0.0000 | 2.4997 | 0.4000 | 27 |
| 28 | 12347.6696 | 0.0001 | 30866.6739 | 0.0000 | 2.4998 | 0.4000 | 28 |
| 29 | 17286.7374 | 0.0001 | 43214.3435 | 0.0000 | 2.4999 | 0.4000 | 29 |
| 30 | 24201.4324 | 0.0000 | 60501.0809 | 0.0000 | 2.4999 | 0.4000 | 30 |

50.00% COMPOUND INTEREST FACTORS

| N | F/P | P/F | F/A | A/F | P/A | A/P | N |
|---|---|---|---|---|---|---|---|
| 1 | 1.5000 | 0.6667 | 1.0000 | 1.0000 | 0.6667 | 1.5000 | 1 |
| 2 | 2.2500 | 0.4444 | 2.5000 | 0.4000 | 1.1111 | 0.9000 | 2 |
| 3 | 3.3750 | 0.2963 | 4.7500 | 0.2105 | 1.4074 | 0.7105 | 3 |
| 4 | 5.0625 | 0.1975 | 8.1250 | 0.1231 | 1.6049 | 0.6231 | 4 |
| 5 | 7.5938 | 0.1317 | 13.1875 | 0.0758 | 1.7366 | 0.5758 | 5 |
| 6 | 11.3906 | 0.0878 | 20.7813 | 0.0481 | 1.8244 | 0.5481 | 6 |
| 7 | 17.0859 | 0.0585 | 32.1719 | 0.0311 | 1.8829 | 0.5311 | 7 |
| 8 | 25.6289 | 0.0390 | 49.2578 | 0.0203 | 1.9220 | 0.5203 | 8 |
| 9 | 38.4434 | 0.0260 | 74.8867 | 0.0134 | 1.9480 | 0.5134 | 9 |
| 10 | 57.6650 | 0.0173 | 113.3301 | 0.0088 | 1.9653 | 0.5088 | 10 |
| 11 | 86.4976 | 0.0116 | 170.9951 | 0.0058 | 1.9769 | 0.5058 | 11 |
| 12 | 129.7463 | 0.0077 | 257.4927 | 0.0039 | 1.9846 | 0.5039 | 12 |
| 13 | 194.6195 | 0.0051 | 387.2390 | 0.0026 | 1.9897 | 0.5026 | 13 |
| 14 | 291.9293 | 0.0034 | 581.8585 | 0.0017 | 1.9931 | 0.5017 | 14 |
| 15 | 437.8939 | 0.0023 | 873.7878 | 0.0011 | 1.9954 | 0.5011 | 15 |
| 16 | 656.8408 | 0.0015 | 1311.6817 | 0.0008 | 1.9970 | 0.5008 | 16 |
| 17 | 985.2613 | 0.0010 | 1968.5225 | 0.0005 | 1.9980 | 0.5005 | 17 |
| 18 | 1477.8919 | 0.0007 | 2953.7838 | 0.0003 | 1.9986 | 0.5003 | 18 |
| 19 | 2216.8378 | 0.0005 | 4431.6756 | 0.0002 | 1.9991 | 0.5002 | 19 |
| 20 | 3325.2567 | 0.0003 | 6648.5135 | 0.0002 | 1.9994 | 0.5002 | 20 |
| 21 | 4987.8851 | 0.0002 | 9973.7702 | 0.0001 | 1.9996 | 0.5001 | 21 |
| 22 | 7481.8276 | 0.0001 | 14961.6553 | 0.0001 | 1.9997 | 0.5001 | 22 |
| 23 | 11222.7415 | 0.0001 | 22443.4829 | 0.0000 | 1.9998 | 0.5000 | 23 |
| 24 | 16834.1122 | 0.0001 | 33666.2244 | 0.0000 | 1.9999 | 0.5000 | 24 |
| 25 | 25251.1683 | 0.0000 | 50500.3366 | 0.0000 | 1.9999 | 0.5000 | 25 |

# Appendix 3
## Effective Versus Nominal Annual Rates

| | Compounding Frequency | | | | | | |
|---|---|---|---|---|---|---|---|
| | Semiannually | Quarterly | Monthly | Weekly | Daily | Continuously | |
| r (%) | $\left(1+\frac{r}{2}\right)^2 - 1$ | $\left(1+\frac{r}{4}\right)^4 - 1$ | $\left(1+\frac{r}{12}\right)^{12} - 1$ | $\left(1+\frac{r}{52}\right)^{52} - 1$ | $\left(1+\frac{r}{365}\right)^{365} - 1$ | $\left(1+\frac{r}{\infty}\right)^{\infty} = e^r - 1$ | r (%) |
| 1 | 1.0025 | 1.0038 | 1.0046 | 1.0049 | 1.0050 | 1.0050 | 1 |
| 2 | 2.0100 | 2.0151 | 2.0184 | 2.0197 | 2.0200 | 2.0201 | 2 |
| 3 | 3.0225 | 3.0339 | 3.0416 | 3.0444 | 3.0451 | 3.0455 | 3 |
| 4 | 4.0400 | 4.0604 | 4.0741 | 4.0793 | 4.0805 | 4.0811 | 4 |
| 5 | 5.0625 | 5.0945 | 5.1161 | 5.1244 | 5.1261 | 5.1271 | 5 |
| 6 | 6.0900 | 6.1364 | 6.1678 | 6.1797 | 6.1799 | 6.1837 | 6 |
| 7 | 7.1225 | 7.1859 | 7.2290 | 7.2455 | 7.2469 | 7.2508 | 7 |
| 8 | 8.1600 | 8.2432 | 8.2999 | 8.3217 | 8.3246 | 8.3287 | 8 |
| 9 | 9.2025 | 9.3083 | 9.3807 | 9.4085 | 9.4132 | 9.4174 | 9 |
| 10 | 10.2500 | 10.3813 | 10.4713 | 10.5060 | 10.5126 | 10.5171 | 10 |
| 11 | 11.3025 | 11.4621 | 11.5718 | 11.6144 | 11.6231 | 11.6278 | 11 |
| 12 | 12.3600 | 12.5509 | 12.6825 | 12.7336 | 12.7447 | 12.7497 | 12 |
| 13 | 13.4225 | 13.6476 | 13.8032 | 13.8644 | 13.8775 | 13.8828 | 13 |
| 14 | 14.4900 | 14.7523 | 14.9341 | 15.0057 | 15.0217 | 15.0274 | 14 |
| 15 | 15.5625 | 15.8650 | 16.0755 | 16.1582 | 16.1773 | 16.1834 | 15 |

| | | | | | | | |
|---|---|---|---|---|---|---|---|
| 16 | 16.6400 | 16.9859 | 17.2270 | 17.3221 | 17.3446 | 17.3511 | 16 |
| 17 | 17.7225 | 18.1148 | 18.3891 | 18.4974 | 18.5235 | 18.5305 | 17 |
| 18 | 18.8100 | 19.2517 | 19.5618 | 19.6843 | 19.7142 | 19.7217 | 18 |
| 19 | 19.9025 | 20.3971 | 20.7451 | 20.8828 | 20.9169 | 20.9250 | 19 |
| 20 | 21.0000 | 21.5506 | 21.9390 | 22.0931 | 22.1316 | 22.1403 | 20 |
| 21 | 22.1025 | 22.7124 | 23.1439 | 23.3153 | 23.3584 | 23.3678 | 21 |
| 22 | 23.2100 | 23.8825 | 24.3596 | 24.5494 | 24.5976 | 24.6077 | 22 |
| 23 | 24.3225 | 25.0609 | 25.5863 | 25.7957 | 25.8492 | 25.8600 | 23 |
| 24 | 25.4400 | 26.2477 | 26.8242 | 27.0542 | 27.1133 | 27.1249 | 24 |
| 25 | 26.5625 | 27.4429 | 28.0731 | 28.3250 | 28.3901 | 28.4025 | 25 |
| 26 | 27.6900 | 28.6466 | 29.3333 | 29.6090 | 29.6796 | 29.6930 | 26 |
| 27 | 28.8225 | 29.8588 | 30.6050 | 30.9049 | 30.9821 | 30.9964 | 27 |
| 28 | 29.9600 | 31.0796 | 31.8880 | 32.2135 | 32.2976 | 32.3130 | 28 |
| 29 | 31.1025 | 32.3089 | 33.1826 | 33.5350 | 33.6264 | 33.6428 | 29 |
| 30 | 32.2500 | 33.5469 | 34.4889 | 34.8693 | 34.9684 | 34.9859 | 30 |

# Index